Historical Maps of Europe

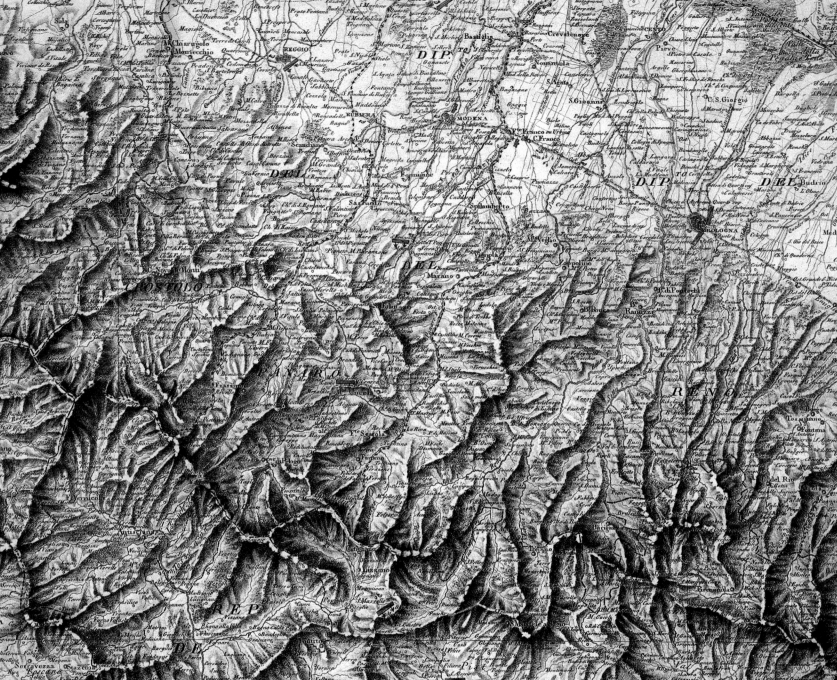

Historical Maps of Europe

Michael Swift

PRC

Acknowledgments

The Public Records Office was the source of all the maps used in this book.

Crown Copyright material in the Public Record Office is reproduced by permission of the Controller of Her Majesty's Stationery Office.

Images are reproduced by courtesy of the Public Record Office.

This edition first published in 2000 by
PRC Publishing Ltd,
Kiln House, 210 New Kings Road, London SW6 4NZ

ISBN 1 85648 575 7

Printed and bound in China

contents

Introduction   **6**

The Dark and Middle Ages: the British Isles 7
The Dark and Middle Ages: Scandinavia 8
The Dark and Middle Ages: Germany 8
The Dark and Middle Ages: France 8
The Dark and Middle Ages: the Iberian Peninsula . .9
The Dark and Middle Ages: Italy 9
The Dark and Middle Ages: the Balkans 9
The Crusades 10
The Fall of the Byzantine Empire 10
The Hanseatic League 11
The Dawn of the Modern Age 11
Europe in the 16th and 17th centuries 12
The Age of the Sun King 13
Europe under the Ancien Regime 13
The Rise of Napoleon 15
The Napoleonic Wars 16
Europe after Napoleon 18
1848 — Year of Revolutions 19

The Cartography

A Note on the Maps 22
Reasons for Cartography 22
The Development of Cartography 23

The Maps

Battle of Mohács 1526 24
Siege of Vienna 1529 24
Ireland 1567 25
Westminster 1585 .26
Cadiz Harbour 1587 27
Wales 1610 28
Western France 1636 29
Portugal 1653 30
Dutch Coast 1673 .31
Tangier 1679 32
Battle of Fleurus 1690 33
St. Malo 1693 34
St. Malo 1700 35

Île de Ré 1700 36
Ghent Canal 1696 .37
Belgium and the Netherlands 1701–12 38
Blenheim 1704 39
Pas de Calais 1707 40
The Rhine 1711 41
German Empire 1715 42
Corfu 1716 44
Belgrade 1717 45
La Rochelle 1718 .46
Aragon 1719 47
Toulouse 1720 48
Low Countries 1730 49
Eastern Lombardy 1736 50
Siege of Philipsbourg 1734 52
Ukraine 1736 53
Mediterranean Sea 1738 54
Prague 1742 55
Finland 1743 56
Scotland 1745/46 .58
Brussels 1745 59
Rome 1748 60
Berlin 1757 62
France 1756 63
Central Europe 1763 66
Partition of Poland 1658–1815 68
Home Counties, England 1774–76 70
River Carente 71
Switzerland 1782 .72
Kingdom of the Two Sicilies 1783 73
Europe 1783–1807 74
Madrid 1785 78
Norway and Sweden 1785 79
Austrian Netherlands 1789 80
Portugal 1790 82
France 1790–91 83
Italy 1803 88
Luxembourg City 1801 89
Europe 1800 90

Europe 1805 102
North Italy 1798 103
Copenhagen 1807 .104
Pyrenees–Adriatic 1811 105
Belgium 1814 106
Spain 1770 108
Peninsular War 1808–14 109
Europe 1815/16 110
Russia 1815/16 111
United Kingdom 1815/16 112
France 1815–16 113
European Courier Routes c.1820 114
Italy 1825 116
Greece, Albania 1827 118
Zante 1821 119
Liverpool–Manchester Railway c.1826 120
Canals of Lancashire c.1826 120
Lancashire c.1820 .121
Montenegro 1856–78 122
Istanbul 1836 123
Russia 1828 124
Black Sea 1848 125
Koblenz 1834 126
Edinburgh 1837 127
Edinburgh Castle 1836 127
Ottoman Empire 1843 128
Hungary c.1845 129
Berlin c.1850 130
Naples c.1850 131
Italy 1853 132
Finland 1854–60 133
Denmark 1864 134
Central Europe 1866–67 136
Europe 1858 138
Sweden 1866 140
Corinth Canal 1893 141
Balkans 1897 142
Channel Islands 143

index of maps   **144**

INTRODUCTION

The continent of Europe stretches from the Atlantic Ocean in the west to the Ural Mountains (in Russia) in the east, and from the Mediterranean in the south to the Arctic Sea in the north. Until the end of the Ice Age, much of modern Europe — in particular the bulk of the British Isles, Scandinavia, Germany and the Baltic states — were covered by ice, and many of the geographical features that will be evident through the maps illustrated in this book are the result of the action of the ice, either through glaciation or as a result of flooding following the rise in temperatures. Another important factor to note is that, up until the Middle Ages when the Ottoman Empire was finally restrained, the history of Europe was marked by an almost constant migration of peoples westwards.

Although many of the earliest civilisations — such as those of Mesopotamia and later Egypt — fall outside the areas traditionally known as Europe, these did have an important role in the fostering of European civilisation. In an age before land travel was easy, seas and rivers formed the essential lines of communication and, of these, the Mediterranean was to become the most important for European civilisation. It is no coincidence that many of the great civilisations of the pre-Christian age were centred around the Mediterranean. In terms of the story of European civilisation, it was the Greeks who initially proved to be dominant. Spreading out from a number of states, Greek civilisation was to have a profound influence on the coast of Asia Minor — through settlements such as the famous city of Troy — in North Africa and in Italy. There is evidence of commercial maritime traffic linking ancient Greece not only with the satellite communities established around the Mediterranean, but also further afield, in the British Isles and elsewhere.

In time, however, the Greeks were to be overshadowed and supplanted by the Romans. The civilisation that was to emerge from central Italy was but one of a number of competing states that existed from the 10th and 9th centuries BC. Initially Rome itself was controlled by another of these states, Etrusca, and Rome's rise to prominence did not start until the 6th century BC. Emerging as an independent entity, Roman power was initially extended through central Italy. In the 3rd century BC Roman power was extended to southern Italy, supplanting the existing Greek supremacy. To the south of Italy, another Greek-established state, Carthage, existed on the North African coast. After a period of conflict — including the famous exploits of Hannibal taking elephants across the Alps to invade northern Italy — Carthage was finally to be defeated in 149BC at the end of the Third Punic War. This victory was to bring Rome its first overseas conquests: Corsica, Sardinia and Spain (Sicily having been gained the previous century during an earlier conflict with Carthage) and from these Roman control expanded. Cisalpine Gaul — the area of northern Italy adjacent to the Alps — and much of Greece passed to Roman control, as did Asia Minor, Cyprus and Crete in this period. Under Julius Caesar Roman control was extended over the whole of Gaul (modern-day France) and the first Roman invasion of the British Isles occurred in 54BC. Caesar also attempted to invade Germany, but Roman power never held sway across the Rhine and for the whole period of the Roman Empire, the border with the Germanic tribes was to prove troublesome. Egypt passed to Roman control as a result of the civil war following Julius Caesar's assassination.

Rome passed from being a republic to an empire under Augustus Caesar in 27BC. For the remainder of its existence, Rome was to be ruled by emperors; these were figures of differing standards of probity and success. Some of their names — such as Tiberius, Nero and Caligula — are amongst the most infamous in history; others — such as Hadrian, Trajan, Diocletian and Constantine — proved to be highly effective rulers of the greatest empire the world had ever seen. Under Augustus Caesar, the empire expanded in the Balkans and in Asia Minor, whilst under his immediate successors, much of the British Isles and the Caucasus passed into Roman domination.

Increasingly, during the first centuries of the Christian era, Rome came under pressure from the tribes in the areas surrounding the empire. Most notably the threat came from the various tribes of Germania — the Vandals, the Huns and the Goths. Gradually Rome retreated from some of its more far-flung provinces — Britannia, for example, was evacuated during the early years of the 5th century AD. However, retrenchment was not to secure the empire's survival; successive waves of barbarian invasions finally brought Rome to its knees, and the last emperor of the west, Romulus Augustus, was defeated by Odoacer in AD476.

Although Rome itself was no more the centre of an empire, Roman civilisation was to continue for a further thousand years through the influence of Constantinople, the new capital established by Constantine in the early years of the 4th century AD. The Byzantine Empire dominated the Balkans, much of Asia Minor and Egypt — as well as Rome and Italy in the period up to AD751 — before the threat of another great empire, the Ottoman, was to lead to its destruction.

The Dark and Middle Ages: the British Isles

Following the withdrawal of Roman forces in AD409, Britannia — effectively England and Wales — suffered from incursions by the Celts of both Scotland and Ireland (neither of which had been conquered by Rome). To defend themselves the British invited mercenaries from Europe. These tribes, largely Anglo-Saxon, initially provided a bulwark. However, Anglo-Saxon support itself gradually turned to invasion and settlement, with the result that England was controlled by a patchwork of Anglo-Saxon kingdoms by the mid-5th century. In Scotland a similar movement saw Irish tribes settle in the west of the country and Anglo-Saxons in the south. The native Celtic tribes were either subjugated or gradually forced westwards.

During the succeeding centuries, there was a gradual emergence of unified kingdoms in England, Wales, Scotland and Ireland, but there remained a continuing threat of invasion, particularly from the Vikings of Scandinavia. The same old pattern unfolded: first raiding and then the establishment of permanent settlements: Dublin, in Ireland, was a major Viking centre from the mid-9th century. In England, war with the most powerful Anglo-Saxon monarchy (Wessex) led to the creation of an area of Scandinavian dominance, the Danelaw, from the late 9th century. Indeed for a period — 1016-1042 — England and Denmark shared a monarch.

In 1042 Edward the Confessor came to the English throne; at his death in 1066, Duke William of Normandy invaded and defeated the English king, Harold, at the Battle of Hastings. This was the start of the Norman conquest of the British Isles. Eventually Norman power would extend over all of England and, during the succeeding centuries, Wales and Ireland. It was also the start of England's close dynastic ties with France. At its height, in the late 12th century, the Plantagenet successors of the Norman dynasty would rule an empire that encompassed much of the British Isles and much of France. For the next four centuries — until England lost Calais in 1558 — there would be almost constant war or threat of war between England and France. Indeed, it was not until the early 19th century that the British monarch finally abandoned all claims to the French throne.

The Middle Ages gradually saw English pre-eminence extended through the British Isles. It was during the reign of Henry II that one of the warring factions in Ireland invited Henry's army over (in 1171), thus starting the long involvement of England (and later Great Britain) in the history of that troubled island. The 13th century also saw greater efforts towards the consolidation of English control over both Wales and Scotland. The dominant figure in the campaign against both nations is Edward I, king 1272-1307. Nicknamed 'Hammer of the Scots', Edward's armies brought English victory in Wales and significant advances in Scotland. It was in 1301 that Edward I declared that his eldest son was 'Prince of Wales'.

North of the border, in Scotland, English power had waxed and waned since the Norman conquest. Nominally independent, Scottish king William I had sworn an oath of fealty to the English king in 1174, but Anglo-Scottish relations were marked more by border skirmishes and war than by peaceful co-existence. Earlier in the 12th century King David I of Scotland had endeavoured to conquer the English counties of Cumberland, Westmoreland and Northumberland, but was defeated and killed at the Battle of the Standards. It was only at the Treaty of York, in 1237, that the border as delineated today came into existence. Later in the 13th century, following the death of Queen Margaret, there were 13 claimants to the Scottish throne. King Edward I of England was invited to adjudge which would succeed; he selected John de Balliol, who responded by swearing an oath of fealty to the English king. However, relations between Edward and John deteriorated and the latter abdicated. Edward assumed the Scottish crown and undertook a major military campaign throughout Scotland in 1296. It was at this time that the famous Stone of Scone was pillaged and sent to England.

The Scots, rejecting the power of Edward I, rebelled; one of the leaders of this rebellion was William Wallace, who was eventually captured and executed in 1305. Despite this setback, a resurgence of Scottish power under a new leader, Robert the Bruce, led to victory over the English at Bannockburn in 1314. The Bruce established himself as king of Scotland and, by the time he died in 1329, he was acknowledged by both the Papacy and the English. Following his death, however, Edward III attempted to re-establish English pre-eminence and it was only with the first of the Stuart monarchs, Robert II, in 1371 that stability was again restored.

The 15th century in England was to be marked by destructive civil war, the War of the Roses, which culminated in the rise of a new dynasty, the Tudors. It was under Henry VIII (king 1509-47) that England broke with Rome. His second daughter, Elizabeth I, ruled 1558-1603. On her death, the crowns of both England and Scotland were united in the person of James VI of Scotland, the first of England's Stuart monarchs.

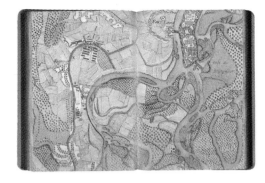

The Dark and Middle Ages: Scandinavia

Never conquered by Rome, Scandinavia was to prove a major influence of the development of Europe in the post-Roman era through the seafaring and military skills of the Vikings. Under the Vikings, the influence of Scandinavia was to spread to the British Isles, northern France, Iceland, Greenland, the Baltic states, western Russia and the Mediterranean.

Denmark was the dominant force through much of this period. It was the Danish-led Union of Kalmar (1397–1523) that saw Norway, Denmark and Sweden united. Danish power reached its zenith in the early 17th century; thereafter war with Sweden gradually saw the latter achieve dominance. Denmark, however, retained sovereignty over Norway until the early 19th century.

For much of the Middle Ages and into the post-reformation era, Finland was disputed territory between Sweden and Russia — it passed to Russian sovereignty in 1807. Close Norwegian ties with Denmark were reinforced through the Union of Kalmar and Norway was to remain under Danish rule even after the dissolution of the union. Sweden was, prior to the union, influential in the spread of Viking power in the eastern Baltic. The Union broke up following a revolt led by Gustavus Vasa, who became monarch; under his successors Swedish territory expanded considerably — the Danes were ejected from their remaining territories in southern Sweden and then Swedish power expanded to the southern Baltic states. The Swedish empire reached its peak after the Treaty of Westphalia in 1648 but was to be destroyed during the Northern War of 1700–21.

The Dark and Middle Ages: Germany

The area occupied by the modern Germany was not occupied by the Romans, although there were regular periods of Roman incursion across the Rhine — the effective border between Roman-occupied Europe and the area outside direct Roman control. As the Roman empire weakened and as the barbarian tribes spread westwards, so a fragmented patchwork of states emerged. Initially, there were eight Germanic kingdoms immediately after the collapse of Roman power, but these were conquered by

Charlemagne in the 8th century as he constructed the Holy Roman Empire. The German states — of which some 400 existed at some stage during the era of the Holy Roman Empire's control of the region (982-1806) — were all relatively weak, unable to resist domination by the Habsburgs from 1438 onwards. The crucial state in Germany was gradually to emerge as Brandenburg-Prussia from the mid-17th century onwards, following the Thirty Years' War.

Unlike Germany, Austria had been conquered by the Romans, but following the collapse of Rome, was settled by the barbarian tribes from the east. Following conquest by Charlemagne, further invaders from the east came in the form of the Magyars, but these were decisively beaten at the Battle of Lechfeld in 955 by Otto the Great. Otto established the Babenberg dynasty, which was to rule Austria until 1246. In 1282, Rudolf I, Count of Habsburg, established his family as rulers; this dynasty was to dominate Austria, and much of Europe, through until the 20th century. The first Habsburg to be declared Holy Roman Emperor was Frederick III, who ruled between 1452 and 1493. Austria was, during the late Middle Ages, the bulwark against the incursion of the Ottoman Empire from the Balkans; indeed Vienna withstood two sieges, in 1529 and 1683, when Turkish ambition sought to extend its influence into central Europe.

The Dark and Middle Ages: France

After the period of Roman occupation, when the area was known as Gaul, the region was, from the early 4th century AD onwards, invaded progressively from various Germanic tribes from the east (the Goths and Franks in particular). Gradually the Franks came to prominence, under the Merovingian dynasty from the mid-5th century AD. One of the most important of the Merovingian monarchs was Clovis (465-511). The power of the Merovingian monarchs was, however, gradually usurped by the 'mayors of the Palace', of whom the most famous was Charles Martel (ruler from 714-741), who was responsible for an epic victory over the Saracens after their invasion from Spain. Another of the 'mayors' was Charlemagne, who deposed the last of the Merovingians in 751 and assumed the throne for himself. Charlemagne established an empire covering much of Europe, becoming Holy Roman Emperor in 800, the date from which the empire is normally assumed to have begun. Charlemagne died in 814. France was to suffer further assaults from the Vikings, who established themselves in Normandy, with the Capetian dynasty being established in 987. The Middle Ages in France were marked by dynastic struggles involving the French monarch against his powerful (and nominal) vassals in the dukes of Normandy, Brittany and Burgundy. The English monarch, following the invasion of England by the Normans in 1066, also had claims to the French throne. These dynastic struggles culminated in the Hundred Years War. France only started to emerge as the unified state once the English and Burgundian threat largely disappeared at the end of the 15th century.

The Dark and Middle Ages: The Iberian Peninsula

Settled initially by Celtic tribes and then conquered by the Romans as their province of Lusitania (along with Spain), the future Portugal was occupied by both Goths and Moors in the period following the collapse of the Roman Empire. It was not until 1179, when the Pope recognised Alfonso I as king, that the country started to emerge. The Moors were finally expelled in 1249 after which internecine war with the country's larger neighbour, Castile, dominated until a peace was established in 1411. From 1385 until 1580 the Avis dynasty ruled the country; it was during this period that the great Portuguese empire was gradually established. After the fall of this dynasty, the Portuguese crown was inherited by Philip II of Spain by force. The Spanish were destined to rule until 1640, when the Braganza dynasty came to power following a nationalistic uprising.

Spain, like Portugal, part of the Roman province of Lusitania, was settled by Visigoths following the Roman collapse. However, the Moors from North Africa invaded in the early years of the 8th century and for the next 300 years much of Spain was under Muslim control. The surviving Christian monarchies were small and relatively weak. Two northern monarchies, Castile and Aragon, were the most significant and it was the increasing power of these that led to the gradual elimination of Moorish power from the peninsula. By 1248 only Granada remained in Moorish hands and, following the union of Aragon with Castile in 1479, even this was lost in 1492. Once the domestic power base was secured, Spain was to emerge as one of the dominant forces in the creation of the New World

and, as part of the Habsburg sphere of influence, in much of Europe. It was Philip II of Spain who launched the ill-fated invasion of England — the Armada — in 1588.

The Dark and Middle Ages: Italy

Although conquered by the Germanic tribes in the 5th century AD, the Italian peninsula was to remain under the control of the Byzantine Empire until AD751. The northern part of the peninsula was to be conquered by Charlemagne in AD775 and was to form part of the Holy Roman Empire from its inception in AD800. The southern half of the peninsula was, at this stage, still controlled by the Byzantine empire, but was being contested by the Arab conquerors of Sicily. Unlike France and Britain, where the Middle Ages gradually witnessed the emergence of single states, northern Italy during this period was to see the growth of powerful city-states, such as Venice, whilst much of central Italy was to be under the control, both spiritually and temporally, of the Pope. South of the papal state came a more unified country. This had, initially, been part of the greater Norman empire but was later, in 1176, to fall under Spanish control. By the start of the 16th century, the peninsula was split into the Habsburg-controlled Kingdom of Naples, the Papal States, Florence, Venice (with control over Dalmatia and many of the islands off the Balkans as a reflection of its trading power), Milan and Savoy. Much of northern Italy — Florence, Savoy and Milan — still formed part of the Holy Roman Empire.

The Dark and Middle Ages: The Balkans

With the creation of the Eastern Empire, centred on Constantinople, the Balkans were eventually to form part of the Byzantine Empire. However, as the empire's power waned in the face of the increasing external threat — latterly from the Ottoman Turks — so control over the Balkan region was reduced. The Middle Ages were to witness an almost inexorable increase in the power of the Ottoman Empire, culminating in the conquest not only of the Balkans but also the fall of Constantinople itself.

The Crusades

Christian domination of the Holy Land was to come under threat from the rise of a new power — that of Islam. The Crusader movement resulted from the conquest of Jerusalem and Syria by the Turkish leader Alp Arslan. Although there were plans as early as 1074 when Pope Gregory VII sought to raise an army to retake the Holy Land, it was not until the Synod of Clermont on 26 November 1095 that Pope Urban II convinced the princes and knights of the rectitude of the campaign. There were two intellectual justifications for the war: the belief in pilgrimage and the concept of a just and holy war to remove the heathens from the spiritual centres.

The First Crusade (1096-99) was led by Robert of Normandy — both King Henry II of England and King Philip I of France were at that time excommunicated for various misdemeanours — and was to prove successful. After a number of victories, the Crusaders undertook a five-week siege of Jerusalem, which fell on 15 July 1099. The Norman-led Crusaders established the Kingdom of Jerusalem under Godfrey of Bouillon, entitled 'Protector of the Holy Sepulchre', who was succeeded by his brother Baldwin in 1100. During this period a number of strongholds were constructed by the Normans — the famous Crusader castles — but the kingdom was to disintegrate in 1187, resulting in the creation of a number of smaller Norman-controlled states often at war with each other.

Renewed Turkish ambition, including the conquest of the earldom of Edessa in 1144, led to the Second Crusade (1147-49) which was led by a Hohenstaufen, Conrad III, of the German states, and French King Louis VII. This crusade was less successful, particularly because of tensions between the French and German forces as a result of inter-dynastic machinations.

In 1187 Saladin, the Turkish Sultan, conquered Jerusalem. This resulted in the Third Crusade (1189-92) led initially by Frederick I Barbarossa. Frederick, however, drowned in 1190 shortly after a stunning victory at Iconium. He was succeeded by his son, Duke Frederick of Swabia, who died in 1191, and was replaced by King Richard I (Coeur de Lion) of England and Philip II Augustus of France. These two led a successful siege of Acre but rather than attempt the reconquest of Jerusalem, Richard concluded a treaty with Saladin. This ceded control of a coastal strip from Tyre to Jaffa to the Christians and guaranteed them the right of pilgrimage to Jerusalem but left the remainder of the Holy Land in Turkish control.

The Fall of the Byzantine Empire

At its peak, the Byzantine Empire controlled much of what had formed the eastern provinces of the Roman Empire. However, its boundaries gradually receded: control over Italy was lost in the mid-8th century AD. The creation of the separate ecclesiastical hierarchy occurred in AD867 when Patriarch Photius separated the Orthodox Church from Rome. This split with western Christendom was to have profound consequences since it led to Catholic forces attacking and looting Constantinople during the Fourth Crusade in 1204 — thereby weakening the empire's defences.

Although its empire had by this date contracted significantly, the Byzantine state still extended across much of the Balkans and the western half of Asia Minor. However, as the 13th and 14th centuries progressed, so the empire's borders retreated further. In 1354 the Ottoman Empire achieved its first foothold on European soil when it conquered the Gallipoli peninsula; at this stage the Ottoman forces benefited from the weakness of the Byzantine Empire, wracked by civil war since 1321. Ottoman power in Europe was further augmented in 1362 by the conquest of Adrianople, one of the few remaining major towns under Byzantine rule and from this date the Byzantine emperor was largely dependent on the Turks for his survival. Despite appeals for western support — Manuel II travelled to Rome, Paris and London in 1399-1402 in a fruitless effort to gather allies — Constantinople finally fell to the Turks on 29 May 1453, and the Roman Empire was finally extinguished.

By this date Ottoman power had already been extended through much of the southern Balkans. Much of the Christian resistance in the Balkans was destroyed at the Battle of the Field of Blackbirds — in modern Kosovo — in 1389. The last Christian outpost on the southern coast of the Black Sea in Asia Minor, Trapezus, was conquered in 1461. Ottoman power was extended to Egypt, Syria and Arabia during the period 1512-20.

In 1520 Suleiman II the Magnificent ascended to the Ottoman throne. In 1521 his forces conquered Belgrade. The following year the Knights of St. John surrendered Rhodes to Ottoman rule. In 1526, a decisive victory at the Battle of Mohács (see map page 24) resulted in the Ottoman army laying siege to Vienna in 1529. In 1541 Hungary was to become a province of the Ottoman empire following an agreement to partition it between the Ottoman empire and Austria in 1533.

Although the Ottoman empire was again to threaten Vienna (in 1683), Suleiman's reign was to mark the apogee of Ottoman power in Europe. The Crimea had been incorporated in 1475, Budchak in 1484, Moldavia in 1512, Jedisan in 1526 and Transylvania in 1541 along with Hungary. The power of the Ottomans was reflected in the fact that the Habsburgs agreed to pay an annual tribute at the Peace of Adrianople in 1568. However, the defeat of the Ottoman navy at the battle of Lepanto three years later was evidence of the rise of Spain. The Ottoman position was further weakened by renewed war between it and Austria between 1593 and 1606, settled by the Treaty of Zsitva-Torok in 1606.

The Hanseatic League

Although corporate associations of German merchants had existed since the 11th century, the body that was later to form the Hanseatic League for the control of trade in the Baltic was first established at Wisby in 1161. It was transferred to Lübeck shortly afterwards and this port came to dominate east–west trade through the Baltic.

It was in 1358 that the League 'van der düdeschen Hanse' was established to secure trading advantages. The league was formed in a relatively loose way; there was never a formal constitution and membership altered over the years. Foreign offices were established in a number of overseas cities — such as London and Novgorod (St. Petersburg). Although primarily a trading operation, the Hanse did get involved in military activity, when its trading rights were threatened. Its most serious opponent during these early years was Denmark, which controlled the customs duties on all ships passing into the Baltic through the straits between Denmark and the Danish province of Scania (located in southern Sweden). In 1435 the Danes were forced to concede free passage to ships belonging to the league.

However, the league was to decline during the 15th and 16th centuries for a number of reasons. Firstly, the growth of stronger national states in the Baltic — particularly Sweden and Russia; secondly, maritime trade increasingly sought to use ports on the Atlantic seaboard; and, finally, a decline in the local fishing industry. The disappearance of the league was reflected in the closure of its offices overseas, with that in London ceasing to operate in 1598.

The Dawn of the Modern Age

Two distinct, but interrelated, trends at the end of the 15th century and start of the 16th helped to shape modern Europe — the Renaissance and the Reformation. The former, largely to be felt in the arts, sciences and political thought, was to see a great flowering of European culture as evinced by the stunning buildings and works of art that date from the era. As with many other cultural influences, the roots of the Renaissance lay in Italy, but it was to influence profoundly culture throughout the continent.

But it was the second of these movements, the Reformation, that was more influential in terms of shaping the political and ethnic map of Europe. The Reformation was a reaction against the failings of the Catholic church and of the excesses of its leading figures. There had been tensions between church and state in many European countries throughout the Middle Ages, but to the reformers, often relatively minor clerics and thinkers, it was the endemic corruption of the church — exemplified by such things as the sale of indulgences — that was the focus for revolt. The reformers initially sought to restore Catholicism to a simpler form but then began to question some of the basic tenets of the faith, such as the unwillingness to accept translations of the bible from Latin into the vernacular and over such concepts as predestination or free will.

The reformist zeal was strongest in northern Europe and saw the creation of Protestant churches in Scandinavia, Switzerland, much of Germany, Scotland, the Netherlands and England. The rise of Protestantism and the division within Christendom was to become another cause of conflict. No longer was it purely territorial ambition that drove states into war but it was also religious sympathy that could help forge alliances. It was no accident, for example, that England was sympathetic to the French Protestant minority in the 16th and 17th centuries when the country was threatened by the Catholic monarchs of France and Spain.

The close inter-relationship between church and state is no better typified than in England where Henry VIII (king 1509-47), determined to secure a male heir to avoid a repeat of the civil war of the previous century, sought a Papal annulment of his barren marriage. Previously regarded as a staunch Catholic — indeed, called 'Defender of the Faith' by the Pope for his attacks on Protestant reformers — Henry's failure to obtain a divorce led him to break with Rome and establish the Church of England.

Europe in the 16th and 17th centuries

There were few years in these centuries when some part of the continent was not at war, but of all the conflicts, it was the Thirty Years' War which was the most widespread and had the most profound consequences. Its origins lay in Austria, where the new monarch, Ferdinand II, came to power at a time when his predecessors had allowed power to seep away from the absolutist monarchy and also had failed to suppress the Reformation in the lands controlled by the Holy Roman Empire. These points were also true in the German lands controlled by the Empire; in September 1555 the Treaty of Augsburg had created a balance between the independence of the various German states against the desire of the Habsburgs for greater control. This treaty, however, only stifled the potential for tension, particularly when religious freedom was again under threat. In theory, such a war would have been only of significance to Austria and the German states. However, Ferdinand II formed an alliance with Spain, another Habsburg domain, which seemed to threaten France, and thus formed a phase in the ongoing struggle for European hegemony fought out between the dynastic rulers of France and Central Europe. Ferdinand and his Spanish ally, Philip IV, were keen to seen action taken against heretics — as they perceived the Protestants to be — throughout Europe. This action, which threatened the recently established Protestant states of northern Europe, brought France allies in the form of England, Denmark, Holland (the result of the Revolt of the Netherlands in the late 16th century, settled by a 12-year truce in 1609), Sweden and Russia.

The first phase of the war occurred in Bohemia, lasting 1618–20. This culminated in the defeat of Frederick V by the Empire and Bavaria at the Battle of White Mountain on 8 November 1620. Frederick V fled and Ferdinand subdued Bohemia and instituted new constitutions making the Habsburgs hereditary monarchs. The second phase of the conflict was the Palatinate War of 1621–23. In this the Upper Palatinate was conquered by Bavaria and the Rhenish Palatinate conquered jointly by Spain and Bavaria. In the Low Countries, following the 12-year truce, conflict again occurred in Holland between the Dutch and the Spanish. The war in the Netherlands was to continue throughout the period up to 1648.

The next phase was the Danish War. This saw a Danish-led anti-Habsburg coalition invade northwest Germany and other coalition members attack Bohemia from the east. The Danish army, led by Christian IV, was defeated at the Battle of Lutter on 27 August 1626. Following the defeat of those forces attacking Bohemia, the pro-Catholic forces drove the Danes out of northwest Germany in July 1627. The Treaty of Lübeck of May 1629 recognised Denmark's defeat and forced the country out of the conflict. Denmark's role was assumed by Sweden under Gustavus II Adolphus. It was from this point that the power of the Empire was gradually weakened.

Sweden's invasion of Germany came in June–July 1630 when Gustavus landed at Peenemünde. Swedish forces captured Pomerania, Mecklenburg and other small North German provinces. Initially, the major Protestant leaders in Germany, from Saxony and Brandenburg, were reluctant to support Sweden, fearing a new domination, but the invasion of Saxony by the Empire's army forced these states into an alliance.

The Empire's army was defeated at the Battle of Breitenfeld in September 1631 and the Swedish-led forces then proceeded to invade southern Germany, reaching Frankfurt in December of that year. In April 1632, Gustavus's army attacked Bavaria. However, the Swedish army, whilst defeating the Empire's forces at Lützen in November 1632, was to suffer the grievous loss of Gustavus, who was killed during the battle.

Following the death of Gustavus, the Swedish Chancellor (Axel Oxenstierna) established the Heilbronn League in April 1633; this brought together Sweden and all the Protestant German states (except Saxony) but the league was to collapse in late 1634 following its defeat at the Battle of Nördlingen. Saxony signed the Treaty of Prague with the Empire in May 1635; this guaranteed certain rights to Saxony and effectively marked the end of the Thirty Years' War as a civil war in Germany.

On 19 May 1635 France declared war; this was to launch a 13-year conflict between France and its allies and the Habsburgs. Although the grand strategies of these opponents were to dominate the conflict, there were subsidiary wars elsewhere in Europe. In the Netherlands, the Dutch sought to strengthen their position in the ongoing struggle for independence from Spain. In Germany, many of the smaller states continued intermittent and unco-ordinated activity as each sought to ensure its own territorial integrity. Sweden, fast becoming the dominant power in the Baltic, sought to extend its own influence in northern Germany and further east in wars such as the Swedish-Danish War of 1643–45.

From the mid-1640s onwards, negotiations began to settle the war. The result was the Treaty of Westphalia, which was signed on 24 October 1648.

This treaty confirmed that France and Sweden were the effective victors in the war. France gained sovereignty over a number of regions, including Alsace. The independence of the Swiss Confederation, hitherto regarded as part of the Empire, was confirmed. The United Provinces, as the Protestant part of the Netherlands became known, were separated from the Spanish Netherlands and granted independence. Sweden obtained control over West Pomerania and the bishoprics of Bremen and Verden, whilst Brandenburg acquired East Pomerania. Without doubt, the major loser at Westphalia was the Empire; it was forced to concede the independence of the German states and thus failed in its primary objective of creating a Catholic monarchy in Germany.

The Age of the Sun King

The French king Louis XIV (*Le Roi Soleil* — the Sun King) dominated Europe at the end of the 17th and start of the 18th centuries. Coming to power at age 22 in 1661, his primary aim was to extend France's power and influence. Through alliances, initially with the Rhenish states along with Sweden, Poland, Hungary and Turkey, he sought to take on the might of the Holy Roman Empire, which then dominated not only central Europe but also exercised control in Germany, part of the Low Countries and Spain.

Louis' first war, the War of Devolution (1667–68), was fought over Spain. This forced Britain, through the Treaty of Breda, into an alliance with the Netherlands and Sweden. This alliance forced Louis to accept the Treaty of Aix-la-Chapelle (1668). Alliances at this time were shortlived and following the treaty, Louis allied with Britain and Sweden to exact revenge on the Dutch. From 1672 until 1678 the French were at war with the Dutch; this was settled by the Treaty of Nijmegen in which the Dutch lost no territory, but Spain was forced to concede Burgundy to the French.

French power was further extended in 1681 by the annexation of Strasbourg and by the occupation of Luxembourg in 1684. This period of conflict was settled by the Truce of Regensburg in 1684, although anti-French opposition still existed through the League of Augsburg established in 1686. This was followed in 1689 by the creation of the Grand Alliance in response to the French invasion of southern Germany.

Much of the French power at this time was concentrated on land but, like its rival in global ambition, Britain, naval power was also important.

The new French fleet was defeated at the Battle of La Hogue in 1692. This phase of the European wars was concluded by the Treaty of Rijswijk in 1697; by this treaty, Louis lost a certain amount of land gained over the previous decade but retained Strasbourg and Alsace.

The next European crisis came with the War of the Spanish Succession. The ruling Habsburg monarch, Charles II ('The Bewitched'), in Spain had no obvious successor. Rival powers saw this as an opportunity to increase influence through proposing rival solutions to fill this gap. Under the influence of Louis XIV and the Spanish Council of State, Charles II nominated Philip of Anjou as heir. Perceived as a threat to the balance of power in Europe, this solution led to the creation of a second Grand Alliance, involving Britain, Austria, Portugal, Hanover, Holland, Prussia and others. The War of the Spanish Succession — with civil war in Spain, naval action, war in the Netherlands and in Germany — was one of the first to be fought through much of continental Europe. In 1704 Britain seized Gibraltar and over the next decade the Alliance won a number of stunning victories — such as Blenheim in 1704 (see map p39), Turin 1706, Oudenaarde 1708 and Malplaquet 1709. However, by 1709 both sides were militarily and financially exhausted with the result that, despite a state of war existing, little action took place and the war was settled by the Treaty of Utrecht in 1713. This treaty resulted in the dismemberment of the Spanish empire. Spain itself and its overseas colonies passed to Philip of Anjou. Spain's European territories passed to Austria, save for Sicily which was ceded to Savoy. Britain gained sovereignty over Gibraltar, thereby helping to establish the British presence in the Mediterranean and forming an essential part in the country's future naval supremacy. Britain also gained territory in North America and rights over the transatlantic slave trade which were to be of fundamental importance in the creation of Britain's 18th century wealth and North American empire.

Louis XIV died in 1715, shortly after the Treaty of Utrecht. He was succeeded by a five-year old, Louis XV, who would rule France for 59 years.

Europe under the Ancien Regime

At the start of the 18th century the map of Europe was significantly different to that familiar today. In Scandinavia, Denmark was united with Norway and still held control over the whole of the primarily German

provinces of Schleswig and Holstein. Sweden, having lost much of its territory on the southern Baltic coast (the future Estonia, Lithuania and Latvia) in the 17th century, was still united with Finland, although the Russians had gained control of the Karelia area. Between Russia and Germany was the huge state of Poland, which stretched from the Baltic almost to the Black Sea — the fate of Poland was to be one of the dominant features of the 18th century. To the southwest of Poland and controlling the bulk of the Balkans — with the exception of Dalmatia, the Ionian Islands (both of which were under Venetian rule) and Montenegro — was the once-threatening Ottoman Empire.

Italy was a patchwork of small kingdoms, with the Papacy ruling Rome and the Papal States. Sicily was united with southern Italy to form the Kingdom of the Two Sicilies. Switzerland was independent. Central Europe was dominated by the Holy Roman Empire — once characterised as being neither Holy nor Roman, it was certainly not an homogenous empire. Dominated by Austria, it was formed of numerous duchies (as was Italy to the south). Austria, itself part of the Empire, included the Tyrol as well as the area later to be formed into Czechoslovakia. United with Austria, but outside the Empire, was the Kingdom of Hungary. Austrian power also extended to part of the Low Countries (the future Belgium), whilst the Netherlands existed (as the United Provinces) to the north. Spain, Portugal and France existed as separate states. Across the Channel, England and Wales were united as a single entity, while Scotland was linked to the other two through a shared monarch. Union with England would come through an Act of Union in 1707, when the Scottish parliament voted itself out of existence. Ireland was under English domination and shared a common monarch, although it retained its own parliament until 1801.

Apart from the territorial relationships that established the fabric of 18th century politics, there were also dynastic links that helped forge alliances over the next 100 years. Of these the most significant was probably the Bourbon family. Through the Bourbons, the royal houses of France, Spain and the Two Sicilies were interlinked. From 1714, when George Elector of Hanover succeeded to the British throne as George I, there were close connections between Britain and Hanover, and this linkage was a major factor in determining British policy in Europe during the century. The British monarch remained Elector of Hanover until the succession of Queen Victoria in 1837; the link was only broken at that stage because a woman could not succeed to the position of elector.

The 18th century was to see two further trends that both affected the nature of Europe by the end of the period. First, just as the 16th century was marked by the Renaissance, so the 18th century was to see a period known as the Enlightenment, where countries were ruled by 'Enlightened Despotism'. Three of the leading powers — Prussia (under Frederick), Austria (under Maria Theresa) and Russia (under Catherine) — all saw dominant rulers in a position to undertake reforms domestically. Second, the 18th century was also an age of increasing European empire building in the Americas and in India. War between European powers was no longer simply about territorial ambition in Europe but was also about extending power overseas; indeed, several colonial wars were fought — particularly between Britain and France — in the colonies at a time when nominally there was peace in Europe.

The only major European wars during the 18th century prior to the French Revolution were the War of Austrian Succession (1740–48) and the Seven Years' War (1756–63). The former was the result of Frederick the Great's ambitions to expand the power of Prussia. Prussia allied itself with France and France with Bavaria. Following an invasion of Silesia in 1740, the Prussians inflicted defeat on the Austrians at the battles of Mollwitz and Chotusitz. The combined French and Bavarian army advanced on Linz and Prague. The first phase of the War of Austrian Succession was settled by the Treaty of Breslain in 1742 by which Austria surrendered control of Silesia. However, this was to be only a temporary peace as Austria negotiates new alliances with Britain, Savoy and Saxony. Following defeat of the French army at Dettingen in 1743, the Franco-Prussian alliance was renewed and the second phase of the war commenced in 1744. Again, Prussian victories, this time at Soor and Hohenfriedberg, resulted in a further weakening of Austrian power, reflected in the Treaty of Dresden in 1745. The War of Austrian Succession was finally concluded in 1748 by the Treaty of Aix-la-Chapelle, which saw France return the Austrian Netherlands to Habsburg control and minor alterations to Italy.

The tensions reflected by this war were not solved by the Treaty of Aix-la-Chapelle. Within a decade Europe was at war again; this time the Seven Years' War. By the Convention of Westminster, in January 1756, Prussia and Britain agreed a defensive treaty to ensure the security of Hanover. This was followed in May 1756 by a treaty between France and Austria, an alliance soon joined by Russia, Sweden, Saxony and other elements of the Holy Roman Empire. Frederick the Great decided on a pre-emptive strike against Saxony and war resulted. Despite facing overwhelming forces, Frederick was able to defeat his opponents at a number of major battles.

During this period, Britain's policy was to subsidise armies in Europe rather than send its own forces into the field. Thus Britain provided significant amounts of money to bolster the Prussian military effort and also funded Ferdinand of Brunswick, who achieved considerable success, with victories at Krefeld in 1758 and Minden in 1759. The primary focus, however, of British policy was to weaken France in both North America and in India; with these dual aims largely achieved by 1761, the British Prime Minister (William Pitt the Elder) effectively ceased to provide the essential funding that Frederick required. The Seven Years' War was settled by the Peace of Hubertusburg; this resulted in no territorial changes in Europe, but the Treaty of Paris, of February 1763, effectively eliminated French influence in North America.

The three acquisitive empires of central and eastern Europe — Austria, Prussia and Russia — were, however, to engineer one vast change to the European landscape in the last quarter of the 18th century — the elimination of Poland (see map p68-69). As mentioned above, Poland at the start of the century had formed a buffer state between Russia and the central European powers. The Polish state, however, was weak and was heavily influenced by the major powers that surrounded it. The first phase in the Partition of Poland occurred in 1772 when Russia gained territory to the west of Smolensk, Austria gained Galicia and Prussia received the region called West Prussia. This grant thus provided a land link between Prussia itself and the German states. The Second Partition occurred in 1793; this saw Prussia gain the region round Posen and Russia expand considerably westwards, incorporating territory that today forms the Ukraine. The final stage in the dismemberment of Poland came in 1795 when Russia gained the Baltic states (like Lithuania) and southwards. Austria gained the region between Galicia and Warsaw and Prussia saw its territory extend eastwards. The end result of the Partitions of Poland was that Austria and Prussia now shared a common border with Russia.

The Rise of Napoleon

The Ancien Regime was to collapse spectacularly as a result of the French Revolution. During the reign of King Louis XVI, the French monarchy became increasingly unpopular; the lavish lifestyle of the royal family and of the leading aristocrats was in stark contrast to the poverty of many sections of the population and the French ruling class was without doubt either unaware of or incapable of doing anything to ameliorate the conditions of the poor.

By the end of the 18th century, the system bequeathed to France by Louis XIV was in ruins. The country was largely bankrupt as a result of wars both in Europe and overseas; the latter had seen much of the French empire in North America and in India lost to the British. Domestically, the rural economy was in tatters as a result of traditional methods of agriculture and of land ownership. In order to reform the state's finances, the king's leading ministers, Necker and Calonne, had introduced reforms between 1787 and 1789 that hit the aristocracy hard, and brought resistance from those affected. The result was that the king was forced to summon the States General, as the French parliament was known, in 1788.

The impetus behind this was from the aristocracy, who felt that they would be able to regain their traditional power, but in this they were sadly mistaken. The States General did not convene until 5 May 1789 and immediately found itself in a procedural dispute as to how it was to be organised. Traditionally, there were three chambers representing the Three Estates — First, clergy; Second, aristocracy; and, Third, commonalty — but the Third Estate demanded that all three should meet as one. Ultimately, the Third Estate forced the hands of the others two, and on 17 June 1789 a National Assembly was established.

At this time, Louis XVI was still in power, although he was increasingly being forced to concede to the new assembly, which changed its name to 'Constituent Assembly' on 9 July 1789. On 14 July 1789, the defining moment of the French Revolution came with the storming of the Bastille — the royal fortress and prison that commanded the east of Paris. With the fortress stormed, the mob demolished the building stone by stone. Ironically, the prison was almost devoid of inmates and the act was more symptomatic of revolutionary fervour than a triumph of freedom. The next phase in France's revolution was *La Grande Peur* (the Great Fear), a widespread belief amongst the revolutionaries that the forces of reaction were about to try and regain power. With the collapse of order and with the revolutionary leaders lacking the means of controlling the crowd, there was widespread panic through many districts of France. In reality, the forces opposed to the revolution were in no position to regain power.

Reforms continued apace with, for example, the abolition of Feudalism, but the economic crisis worsened. In October 1789, the Bread March of the Women from Paris to Versailles forced the king and his family to return to Paris. For the next two years, France appeared to be moving towards a

constitutional monarchy through the influence of the moderate reformer Mirabeau. However, his death in April 1791 removed one of the restraining forces from the revolutionaries.

A number of factors gradually propelled the revolutionaries towards republicanism. First, despite the limitations on his power, Louis XVI continued to attempt to stop reforms of the church. Second, the family made an ill-fated attempt to escape — the Flight to Varennes — on 20-21 June. Finally, and perhaps most crucially, Queen Marie Antoinette was Austrian. The Austrian monarchy and its German allies were antipathetic towards the revolution and provided sanctuary to a number of emigré aristocrats. On 20 April 1792 the French declared war on Austria and Prussia and, as a result, riots broke out in Paris. On 10 August 1792 the Tuileries Palace was attacked by the mob and Louis XVI and his family imprisoned. The reality of invasion and the fall of Verdun brought increased rioting and a massacre of prisoners who were suspected of being traitors. On 21 September 1792 the Revolutionary Convention formally abolished the monarchy.

The failure of the French army brought internal conflict between the two revolutionary factions, the Girondins and the Jacobins. Initially the former — led by Brissot, Roland, Roland's wife Pétion, and Vergniaurd — held sway. It was under their aegis that Louis XVI was executed (on 21 January 1793) along with other members of the royal family. After initial military success, war was declared on Britain, Holland and Spain, but victory quickly turned to defeat and in mid-1793 the Jacobins assumed power, executing the leading Girondins in October 1793.

The Jacobins were the most radical of the revolutionaries; led by Robespierre, who defeated factions led by Hébert and Danton, who were executed, as were some 2,600 in Paris alone during the period of Jacobin supremacy and many others elsewhere. This was the period known as the 'Terror'. However, the Jacobins were themselves to be overthrown in mid-1794 and Robespierre was to go to the guillotine himself on 28 July 1794.

Following the demise of the Jacobins, there was a 14-month period when moderates held sway. This period ended in October 1795 and was brought to a close both by further rioting in Paris and by a new constitution that allowed for a five-man elected leadership, the 'Directory'. Initially, the Directory, of which Barras and Carnot were the influential figures, was successful, but increasing internecine strife between the leaders, along with anarchy and pro-monarchist risings in the provinces, brought a new coup on 9 November 1799. It brought to power a Corsican soldier who would redraw the map of France and Europe — Napoleon Bonaparte.

The Napoleonic Wars

The French Revolutionary and Napoleonic wars were to involve Europe for more than 20 years — although there would be several periods of peace during that time. The common thread throughout the period was, however, the shifting patterns of the coalitions that fought the war.

As already indicated, the final impetus to the execution of Louis XVI and his family came from the threat of an Austro-Prussian invasion of France and the first phase of the war was, indeed, fought between France and an invading force from Prussia and Austria. The significant battles of this early phase of the war were at Valmy (on 20 September 1792) and at Neerwinden (on 18 March 1793). This first period of the conflict was known as the War of the First Coalition.

The original alliance, between Austria and Prussia, was extended by the addition of Sardinia-Piedmont (1792), Britain, the Netherlands and Spain (in 1793) and later by Naples and the Papal States. Initially, the coalition had successes, with Prussian forces invading France, but this incursion was to be forced back and the French countered with a successful invasion of the Netherlands in 1794-95. In March 1795 the Prussians concluded a separate peace, as did Spain in June of that year. The victorious French then conquered Piedmont, Naples and the Papal States in 1796. Napoleon's victories in Italy resulted in the Treaty of Campo-Formio in October 1797 with Austria. The result of this treaty was that the future Belgium was transferred from Austrian to French sovereignty, whilst France also gained the Ionian Islands. Northern Italy was to form a Cisalpine Republic, whilst Austria was handed control of Venice. Secret clauses in the treaty allowed for the partition of much of the German land between France and Austria.

The results of these various treaties meant that by 1797 Britain was alone against the French. Although the French never got as far as attempting an invasion of mainland Britain, they certainly prepared for one, and coastal defences were constructed around the south coast of England to counter such a threat; the French also made efforts to sow seeds of disaffection in Ireland, where rebellion was never far from the surface.

Theobald Wolfe Tone was one of the leaders of the United Irishmen and it was he, following a period of exile in France, who was to assist the French to invade Ireland at Bantry Bay in December 1796. This major force was, however, largely dispersed by adverse weather conditions and the invasion failed. This was to be one of a number of attempted incursions and rebellions. On 22 August 1798 a further French force under General Humbert landed at Killala; a government army was defeated at 'Races of Castlebar' on 27 August but Humbert was forced to surrender at Ballinamuck on 8 September. Following a further French raid on Lough Swilly, Tone was arrested on 3 November 1798; he committed suicide in prison on 19 November. It was the threat from France that was one of the major factors behind the Act of Union in 1801 which resulted in the dissolution of the Dublin Parliament.

The peace established by 1797 was, however, to be shortlived and war soon reappeared. The War of the Second Coalition lasted from 1798 until 1801. The alliance was forged by the British Prime Minister, William Pitt the Younger. In December 1798 Britain, Austria, Naples, Portugal, Russia and Turkey agreed to act in unison against the French — Spain rallied to the French cause. At this time Napoleon was occupied with his campaign in Egypt, with the Battles of the Pyramids (21 July 1798) and Aboukir Bay (1 August 1798). Defeated in his endeavours, Napoleon returned to France in 1799 where he was to lead a coup against the government (the Directory), which led ultimately to him becoming emperor.

Initially, the allies against Napoleon had some success, particularly in Italy, but the tensions between Russia and Austria led to the former withdrawing from the alliance in October 1799. In January 1800 the Turks made peace and in May 1800 Napoleon led his army against the Austrians with victories at Marengo (14 June 1800) and at Hohenlinden (on 3 December 1800). Defeated, the Austrians agreed to peace at the Treaty of Lunéville in February 1801. The Portuguese were defeated by Spain in September 1801. Britain again remained alone until peace was concluded at the Treaty of Amiens on 25 March 1802. By this treaty Britain retained various colonial acquisitions (such as Ceylon and Trinidad) but returned others (Malta to the Knights of St John, Egypt to Turkey, and so on). The independence of the Ionian Islands and Portugal was guaranteed. The British monarch agreed to be no longer styled 'King of France' — a title claimed since the Middle Ages but untenable following after the death of Henry V. In theory, Amiens was designed to bring lasting peace, but in reality it was to last only 14 months, with war recurring from 16 May 1803.

Again Britain was at war alone; it was not until 1805 that the Third Coalition was established. In April 1805 Russia joined with Britain with Austria following in August of the same year. However, the Third Coalition was to be a failure. France defeated the Austrians at the Battle of Austerlitz

(2 December 1805) and sued for peace at the Treaty of Pressburg (26 December 1805). Earlier, Britain had achieved a famous naval victory at Trafalgar (21 October 1805), when Admiral Lord Nelson had defeated the combined Franco-Spanish fleet.

Although Austria was forced from the coalition, Prussia joined with Britain and Russia in mid-1806, but French victory at Jena and Auerstadt (14 October 1806) forced the Prussians to seek peace. Russian resistance continued until mid-1807 when, in a complete *volte face*, Russia changed sides following the Peace of Tilsit (7–9 July 1807).

Once again left to battle against the French alone, Britain's strategy turned to war against France's ally, Spain. Sir Arthur Wellesley (later the Duke of Wellington) landed at Lisbon in July 1808 with an army to be used in support of the Portuguese and Spanish opponents of Napoleon. The Peninsular War was to last until 1814, with both sides achieving victories. The balance swung decisively against the French after British victories at Salamanca (July 1812) and Vittoria (June 1813). Forced to evacuate Spain, the French army was to suffer further defeat, at Toulouse, in April 1814.

Perhaps the defining moment in the Napoleonic era came in 1812 when the French emperor launched his invasion of Russia. With battles at Smolensk (18 August 1812) and Borodino (17 September 1812), the French achieved victories over the Russians, reaching the gates of Moscow. However, during the winter of 1812–13, the French army was ill-equipped to deal with the conditions. Forced into a retreat, the Russian campaign marked the start of Napoleon's fall from power.

Emboldened by the French defeat in Russia and by Prussia's change of sides, the British Foreign Secretary (Lord Castlereagh), established a new coalition. The Fourth Coalition, with Britain, Russia and Prussia, was formally signed in June 1813. Sweden was also to join, as were Austria (August 1813) and Bavaria (October 1813). The coalition was further strengthened by Württemberg and Saxony, who joined during the Battle of Leipzig (16–23 October 1813). The Saxons started as allies of France during the battle, swapping sides during the actual conflict. Napoleon was forced to retreat across the Rhine and the battle marked the final stage in the liberation of Germany from Napoleonic influence.

Following this victory, coalition forces invaded France and entered Paris on 31 March 1814. Napoleon was forced to abdicate on 11 April and a preliminary peace treaty was signed on 30 May. Napoleon was sent into exile on the island of Elba; however, this exile was destined to be shortlived and he escaped back to France in February 1815. He advanced on Paris and

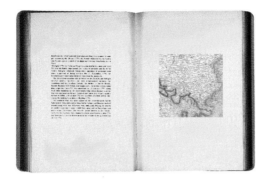

from 20 March to 22 June 1815 he was once again Emperor of France. This period, known as the 'Hundred Days', was to come to an end when the French forces were defeated at Waterloo (18 June 1815) by the British army under Wellington and the Prussians under Blücher. Following this defeat, Napoleon was again sent into exile, this time to the remote British colony of St. Helena in the South Atlantic, where he died in 1821.

Europe after Napoleon

Following Napoleon's defeat at Waterloo, the map of Europe was redrawn as a result of the Congress of Vienna, which had been held between September 1814 and June 1815. The Napoleonic Empire had included all of France, the Low Countries as well as much of Italy; the Congress sought to restore much of the pre-1789 status quo.

The leading players at the Congress were Britain (represented by Lord Castlereagh and the Duke of Wellington), Austria (Metternich), Prussia (Hardenberg) and Russia (Tsar Alexander I, Stein, Nesselrode and Capodistrias). France (represented by Talleyrand) was to achieve a prominence through astute playing off of the various factions. The final settlement, the Treaty of Vienna, was signed on 9 June 1815. The treaty created three new states: the Kingdom of the Netherlands (comprising, Luxembourg, Belgium and the Netherlands), the German Confederation (formed of 39 smaller states) and a free city of Cracow (ultimately to pass to Austria in 1846). Two 'client' states were created: Lombardy-Venetia (whose king was to be the Austrian Emperor) and Poland (ruled by the Russian Tsar).

Elsewhere, the pre-Napoleonic ruling families were restored in Spain, Naples, Modena, Piedmont and Tuscany. The Swiss Confederation and its neutrality was reconfirmed. The major powers, in addition, carved up other territory. Austria gained Dalmatia, Carniola, Salzburg and Galicia. Austria also gained the Swedish provinces of Pomerania, on the south side of the Baltic; Sweden's loss of Pomerania was, however, countered by its acquisition of Norway, which had been united with Denmark. Prussia gained sovereignty over much of Saxony, Westphalia, Posen and Danzig. Britain saw its overseas empire expanded to include Ceylon, Cape of Good Hope, Mauritius, Tobago and Saint Lucia; Britain's maritime role was also reflected in its retention of Malta and Heligoland (returned to Germany in 1890 in exchange for territory in East Africa); Britain was also granted a protectorate over the Ionian Islands, which it retained until 1863.

There were, however, numerous factors which meant that this settlement was inherently weak and throughout the 19th century increasing nationalism and revolution was to lead to further conflict. Many of these problems were to affect the Balkans, the bulk of which remained under the control of the Ottoman Empire. Throughout the century, the Ottoman Empire was widely perceived as the 'Sick Man of Europe' and politics, particularly for the western empires — most notably Britain and France — was how to manage the inexorable decline of Turkey and the increasing threat from Russia.

It was to be in the Balkans that the first real crisis of the post-Napoleonic era occurred. Greece had been part of the Ottoman Empire for some 400 years, but towards the end of the 18th century an increasing cultural awareness (along with the influence of Russian agents) brought Greek nationalism to the fore. In 1821 the War of Independence started with an unsuccessful rebellion in the Danubian Principalities (Moldavia and Wallachia) and in the Morea. Western sympathies were with the Greeks and the fight for independence brought many romantics to the Balkans in support. One of the most famous was the poet Lord Byron, who was to be killed at the Siege of Missolonghi in April 1824. Greek independence had been declared at Epidauros on 13 January 1822 but fighting was to continue until 1829 when Turkey was forced by the leading powers to concede to a new Greek kingdom covering the area south of the Gulf of Volo. Greece was to expand in 1863 through acquisition of the Ionian Islands, in 1881 through the absorption of Thessaly and part of Epirus (from Turkey) and, much later (in 1913), by unification with Crete and the occupation of part of Macedonia and Thrace.

The Danubian Provinces were occupied by Russia 1829-34 and again 1848-51. The territory was again occupied by Russia in 1853 when the Tsar sought to put pressure on the Turks; this was one of the factors that led to Britain and France invading the Crimea. Between 1854 and 1857 the area was occupied by Austria in order to keep the peace, and at the Treaty of Paris in 1856 a devolved administration from the Turkish state was guaranteed; Wallachia and Moldavia united in 1862 to form Romania but it was not until 1877 that the country formally declared independence.

The Kingdom of Belgium was to be created in 1831 following a revolt of the population in August 1830 against the united state created in 1815. In 1839 the Treaty of London guaranteed Belgian neutrality. As part of the Treaty, Belgium was granted part of the western part of the Duchy of Luxembourg, itself part of the united Kingdom of the Netherlands. The remainder of Luxembourg was to remain part of the Netherlands until 1890, with the King of the Netherlands acting also as Duke of Luxembourg. The division in 1890 arose from the fact that while a woman could accede to the throne of the Netherlands, she could not inherit the Duchy of Luxembourg. Thus, over a period of 60 years, three states emerged. Belgium was, and remains, a deeply divided country, with the Dutch-speaking Flemish population and the French-speaking Walloons. Even today, the long-term survival of united Belgium is by no means certain.

1848 — Year of Revolutions

A decade after the Treaty of London, Europe was convulsed by a series of revolutions. Many of these were nationalist in origin, but also owed a great deal to both economic unrest and a desire for greater political reform. The revolutions of 1848 helped to shape the rest of the European century. As in the last century, France was to lead the way. In late February 1848, the Paris mob revolted against the refusal of King Louis Philippe to extend the franchise. The mob forced the king to abdicate, thus bringing to an end the July Monarchy, and allowing for the creation of the Second Republic. This first revolution had left-wing overtones and was led by socialist Louis Blanc. In June, a further revolution saw the suppression of the earlier rebellion. Later in the year Louis Napoleon was elected president. He was the nephew of Napoleon Bonaparte and in December 1851 he staged a *coup d'état* which strengthened his position. In December 1852 he declared himself Emperor (the Second Empire). His reign was to last until 1870, although his position was weakened by an ill-judged attempt to involve France in Mexican politics in the 1860s.

Various parts of the Italian peninsula were also to see rebellion from the early months of 1848. After the settlement of 1815, Italy had been left as a patchwork of states, and much of the north of the peninsula was still in the hands of Austria. Italian nationalism and a desire for political reform were the motivational forces for revolt in Tuscany, Naples, Piedmont and the Papal States. Constitutions were demanded and granted, but these were shortlived. The Piedmontese declared war on Austria on 23 March, following five days of rioting in Milan (which had forced the Austrian commander to withdraw), but lack of support from the other Italian states led to Piedmont's defeat. Rioting in Rome led to the Pope, Pius IX, being forced to flee on 25 November 1848 and the declaration of the Roman

Republic. This was to last until July 1849, when it was suppressed by the French. A restored republic in Venice was to last from March 1848 until August 1849.

Austria itself was to be severely affected. Metternich, one of the architects of the 1815 settlement, was forced to resign following rioting in Vienna in March 1848 and Emperor Ferdinand was forced to flee in May rather than concede increased democracy. In June, following the establishment of a Slav Congress and the accidental killing of the Austrian military commander, a military government was established in Prague; a similar solution was also adopted in Vienna later in the year. Habsburg power was eventually restored, but only after Ferdinand abdicated in favour of his nephew Franz Joseph. Another part of the Austrian empire, Hungary, was also plagued by nationalism. Led by Lajos Kossuth, the Hungarian Diet (parliament) adopted a programme for self-government on 15 March 1848; the March Laws were accepted by Emperor Ferdinand. However, the emperor was able to use Croat resentment at the Hungarians' refusal to accede to their independence to allow a Croat invasion in September of that year. Initially, the Hungarian forces were able to defeat the invading force, allowing Kossuth to claim total independence from Austria; however, by July 1849 the situation had deteriorated. The Hungarians were faced not only by a joint Austro-Croat force, but also by a Russian army offered to Franz Joseph and by rebellions in both Serbia and Transylvania. Defeated in two battles during early August 1849, the brief period of Hungarian independence came to an end. It was not until 1867 that Hungarian self-government was again permitted. This time, Austrian weakness following defeat by Prussia saw Hungarian demands met. From 1867 until 1918, the Habsburg monarchy was known as the Austro-Hungarian Empire.

German nationalism was also to bring revolution in 1848. There were pressures both for political reform and for the creation of a united Germany. Rioting in Berlin during March 1848 forced the Prussian king, William IV, to accept the principle of a united Germany and to call a constituent assembly. The smaller German states generally followed Prussia's lead. It was during the early phases of the revolution that the Frankfurt Parliament was established. This was supposed to represent a provisional parliament of all Germany elected on a universal male suffrage; it was to sit from May 1848 until June 1849. German nationalist sentiment undoubtedly was a factor in a brief war with Denmark over Schleswig-Holstein and in the suppression of a brief pro-Polish uprising in Posen.

While there was a minor rebellion in the Danubian Provinces (see above), the 'Sick Man of Europe' (Turkey) was largely bypassed by the events of 1848. However, strategic considerations meant that both Britain and France became increasingly concerned with Turkey after 1848. The primary factor in this was concern over the growing strength of Russia. Britain had cause to be worried about Russian expansion into India and conflict in Afghanistan. Franco-Russian relations were, at the time, at a low ebb as a result of a dispute over the privileges of Catholic and Orthodox monks in Palestine. For many years Russian foreign policy had been dedicated towards obtaining a warm water port. Controlling much of the coastline of the Black Sea meant that Russia was able to construct a fleet based at Odessa. However, without access to the Mediterranean via the Dardanelles, this fleet was useless. The straits were controlled by the Ottoman Empire and thus Russian ambitions could only be achieved by weakening Turkey. To this end, Russia had been active in support of many of the Balkan nationalist movements.

The immediate cause of the crisis that led to the British and French involvement in the Crimea was a demand, rejected by the Turkish authorities, to allow Russia to act as protector to the Christian minorities within the Ottoman Empire. Turkey declared war on Russia on 23 September 1853. The war was initially concentrated in the Danubian Provinces, although Austrian intervention brought a Russian withdrawal from the area. However, on 30 November 1853 the Turkish fleet was destroyed at Sinope and, in order to prevent a Russian invasion of Turkey, the British and French fleets were despatched to the Black Sea. In March 1854 war broke out between Russia and Britain and France; this was followed in September by British and French troops landing on the Crimean peninsula and commencing a year-long siege of Sevastapol. There were two major set battles during the early months of the war: Balaklava and Inkerman. It was during the former that the famous 'Charge of the Light Brigade' took place. The war brought considerable evidence of military incompetence on the part of the British — this was the war in which Florence Nightingale ('The Lady of the Lamp') radically improved the medical care of the wounded — that led to a complete overhaul of the British army. Despite the addition of 10,000 soldiers from Piedmont, the siege became a stalemate and it was only a threat of Austrian intervention that forced a settlement on the reluctant Russians. A preliminary peace was agreed on 1 February 1856 and a final settlement was signed at Paris a few weeks later.

Although the nationalist sentiment towards both German and Italian unification was thwarted in 1848–49, the belief evinced at the time did not disappear. In both Italy and Germany the second half of the 19th century gradually saw this aspiration achieved, albeit by different means.

The movement towards Italian unification is generally referred to as *Risorgimento* (literally meaning resurrection) — a word that became accepted following the establishment of a newspaper of that name by Count Camillo Cavour in 1847. Cavour was a Piedmontese politician under whose aegis Lombardy was freed from Austrian rule after the war of 1859 and Parma, Modena, Tuscany and the Romagna were united with Piedmont in January 1860. Prior to 1847 Italian nationalism had been largely fostered by a secret society, the *Carbonari* (charcoal burners), that was republican and nationalistic. It had promoted several rebellions in the early 1820s, and by the early 1830s it had been subsumed into the Young Italy grouping led by Giuseppe Mazzini. Again republican in spirit, Mazzini was to be overshadowed in the campaign for unification by one of his supporters, Giuseppe Garibaldi. Mazzini's unwillingness to accept a monarchy meant that he was marginalised, lived in exile and was only able to return to his homeland in disguise just prior to his death in 1872.

Garibaldi is one of the best-known names in Italian history. After time in exile in Argentina, he was influential in the defence of the Roman Republic in 1848–49. In 1859 he supported the Piedmontese in their war against the Austrians, before sailing from Genoa in May 1860 with his Redshirts to invade Sicily and Naples. The forces of the Kingdom of the Two Sicilies were defeated at Calatafimi and on the River Volturno, after which Garibaldi surrendered his newly won territory to the Piedmontese, whose king was declared the first king of the united Italy. After the conquest of the Two Sicilies, Rome was the next target. Garibaldi led two unsuccessful attempts to capture the city (in 1862 and 1867). During the latter campaign, the power of the Vatican was maintained by French forces. Rome was occupied until 1870 by the French; when the Franco-Prussian War required the withdrawal of these troops, Italian forces broke through the walls on 20 September 1870. On 2 October of that year, following a plebiscite, Rome was annexed by Italy; but it would not be until the Lateran Treaty of 1929 that the Papacy accepted the *de facto* loss of Rome and the creation of the city state called the Vatican. The final stage of Italian unification came in 1919 when, after the Treaty of St. Germain, Trentino, Istria and the South Tyrol were added to the state.

If Italian unification was achieved through a combination of populism, represented by Garibaldi, and state power, represented by the ambition of the Piedmontese state, the unification of Germany, despite the aspirations of nationalists, was a triumph of strategy for one individual — Otto von Bismarck, who had been appointed Chief Minister of Prussia in September 1862. His immediate task was to undertake the reform of the Prussian army. Bismarck's belief in national unity was pragmatic; he believed in it, but only if it could be achieved through Prussian dominance. The first phase in his policy was war, allied with Austria, against Denmark over the duchies of Schleswig-Holstein.

The Danish crown had held these duchies since the Middle Ages. However, the population of Holstein was almost entirely German — and the duchy had been incorporated into the loose German Confederation since 1815 — while that of Schleswig was partially German. Danish nationalists were keen to integrate both duchies wholly within the kingdom, but the German population of Holstein, in particular, was opposed to this. A brief war was fought in 1849 over the Danish actions, and a compromise solution was reached in 1852. This was to prove only a temporary settlement. When Danish King Frederick VII died in 1863, he was succeeded by Christian IX, whose right to the duchies was disputed by the German states. They believed that the rightful heir was the Duke of Augustenberg as, under Salic Law (which denied the rights of inheritance through women), Christian was excluded. In a brief war in 1864, the Danes were quickly defeated and, by the Convention of Gastein in 1865, Austria gained the administration of Holstein and Prussia that of Schleswig.

In 1866, following accusations of anti-Prussian activity in Holstein, Bismarck launched a war against the Austrians and the other German states. The Austrian forces were quickly defeated at the Battle of Sadová (near Königgrätz) in the Seven Weeks' War. This victory allowed for the creation of the North German Confederation.

The final phase in the unification of Germany came with the Franco-Prussian War of 1870. The ostensible cause of the war was Prussian support for a Hohenzollern candidate to the Spanish throne, but the immediate cause was French resentment over the Ems telegram. This telegram, sent on 13 July 1870, was the result of an interview between the King of Prussia and the French ambassador, in which the latter had asked for assurances over the Prussian position on the issue of the Spanish suc-

cession. The telegram, as doctored following receipt by Bismarck, implied that both the king and ambassador had insulted the other. Following its leaking to the press, anti-Prussian feeling grew in Paris with the result that the French declared war on 19 July 1870.

France was ill-prepared for war and suffered a number of defeats, culminating in abject surrender at Sedan on 1 September 1870. Both the French emperor, Napoleon III (who subsequently abdicated), and army commander, Marshal Bazaine, were captured. The Germans then lay siege to Paris — a siege that was to last from 19 September 1870 to 28 January 1871. After enormous privation Paris fell, and a peace was agreed on 1 March 1871. The Treaty of Frankfurt of 10 May 1871 recognised the German victory and saw the provinces of Alsace and Lorraine transferred to German sovereignty. Prussian victory also saw the remaining southern states of Germany united with the North German Confederation. In January 1871, Bismarck was able to announce, at Versailles, the creation of the German Empire under the Prussian king as emperor.

With German and Italian unification, the countries of western Europe had, for the most part, arrived at the form in which they were to enter the 20th century. There was still competition between them, but this was reflected more in colonial ambition than territorial aggrandisement in Europe. France harboured resentment over its defeat in 1871, but it would have to wait two generations before it would gain revenge. The only real exception to this position was in Scandinavia, where Norway was still united to Sweden; complete Norwegian independence would not be achieved until 1905.

Further east, however, the European position was much less well defined. Although Turkey was insecure, the Russian empire was also teetering on the edge of rebellion. Perceived as having been defeated in the Crimean War and later in the Russo-Japanese War of 1904-05, the Russian regime during the late 19th century was threatened both domestically and internationally. Repression at home resulted in the rise of radical opposition to the monarchy. Tsar Alexander II, who succeeded to the throne on the death of Nicholas I in 1855, was initially a reformer. Among the policies he adopted was the emancipation of the serfs in 1861 and a new legal code in 1862. However, unrest in Poland during 1866 and an attempted assassination forced the Tsar to adopt more despotic measures. A secret terrorist group, the People's Will, was established and condemned Alexander to death; following several unsuccessful attempts, the Tsar was

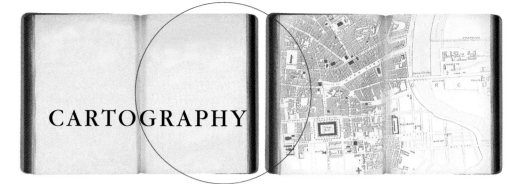

CARTOGRAPHY

killed by a bomb thrown by a Polish student on 13 March 1881. He was succeeded by Alexander III, during whose reign the first Marxist cells in Russia appeared and an alliance with France was established. Alexander III died in 1894, to be succeeded by his son, the last Russian Tsar, Nicholas II.

In January 1905, following humiliating defeat by Japan, revolution began in Russia. The insurrection continued through the year, including a mutiny on board the Black Sea fleet's battleship, the *Potemkin* (an action immortalised in Eisenstein's film). On 30 October 1905, the Tsar conceded a new constitution; this had the effect of splitting the rebels, with the majority accepting the limited reforms on offer. Resistance continued into 1906, and was suppressed by draconian action. The 1905 constitution failed to satisfy the radicals and also failed to solve the inherent weaknesses within the Russian state; it would, however, take war to complete the transformation.

In the Balkans, the Treaty of San Stefano (which ended the Russo-Turkish war of 1877-78) granted formal independence to the part of Serbia formerly controlled by Turkey, along with independence for Bulgaria and confirmation of both Romanian and Montenegran independence. However, this treaty was soon countermanded by the Congress of Berlin (June–July 1878), which reduced the size of Bulgaria. In 1885 Bulgaria, ignoring the wishes of the leading European powers, went to war with and defeated Serbia. The Balkans were a patchwork of nationalities and religions (see map p142), and throughout the late 19th century there was an almost constant threat of war.

In March 1912, at Russian instigation, Serbia and Bulgaria formed an alliance to partition Macedonia, then still a Turkish province. Greece and Montenegro then joined the alliance and, in October 1912, the quartet launched an attack on Turkey. The great powers succeeded in achieving peace of a kind when, in 1913, the Turks agreed to withdraw from most of their Balkan territory, provided an independent Albania was created. Serbia and Greece acquired Macedonia, much to the opposition of Bulgaria, which then launched an offensive on 29 June 1913 against its erstwhile allies. Bulgaria was rapidly defeated and, at the Treaty of Bucharest in August 1913, was forced to concede Macedonia to Serbia and Greece, and Dobrudja to Romania. The Balkan Wars achieved little other than confirming the instability of the region; instability that was to have fatal consequences for the whole of Europe the following year. But that is where we must leave the history of Europe, for this book looks only as far as the First World War; the climactic history of 20th century will be covered in later book.

A Note on the Maps

All the maps illustrated in this book have been drawn from the large collection held by the Public Record Office at Kew in west London. This is the major holding of all public documents in the United Kingdom. The maps are derived from a number of government departments and reflect the interests and concerns at the time they were compiled. These maps were not just produced by British cartographers: there was a general European interest in mapping the continent and a general market in maps.

Reasons for Cartography

At a most basic level maps were required to record the physical presence of land and of physical features. In particular, information regarding safe havens was required and, as knowledge increased, so these charts developed into maritime charts detailing safe channels, expected water depths, hazards to navigation and so on. It is important to remember that for much of the period covered by these maps, all trade was sea-borne and the vulnerable ships of the period were at the mercy of the harsh climate often to be encountered.

Out of discovery came possession and many of the maps that were produced from the mid-16th century were designed to show land ownership. On a global scale, these maps could illustrate the distribution of land between the great nation states, often as a result of a settlement after a war. At a more local level, the maps could illustrate the ownership of parcels of land that had been divided between individual landowners. Whilst, certainly in the early days, the more global representation of the region was suspect, the smaller the areas covered the greater the accuracy. There had been a long tradition of detailed estate maps in Europe, in particular amongst the abbeys and major land owners, and mapping of a small locality was, therefore, a skill widely practised.

Finally, from possession comes conflict. The military were among the most important map makers, with the skills and resources to undertake precise surveys. It is no accident that the United Kingdom's primary mapping agency has the name Ordnance Survey, as it grew out of a department of the military.

The Development of Cartography

After the demise of the Roman Empire, European culture had, to a significant extent, lost many of the skills and arts that the classical civilisations of Egypt, Greece and Rome had possessed. Among the skills that disappeared during the so-called Dark Ages was cartography. The Greeks and Romans had had the skills and the knowledge to produce quite detailed maps which bore some resemblance to the actual landscape and topographical details; post-Rome, however, the civilisations of western Europe lacked the cartographers with the knowledge to undertake the work. Religion had a great deal to do with this. The famous *Mappa Mundi*, on display at Hereford Cathedral in Britain, was completed by Richard of Haldingham in the 13th century; this map, purporting to show the whole world, has at its centre Jerusalem, reflecting the contemporary Christian belief that the earth was flat, the sky represented the heavens and that all centred on Jerusalem.

The Renaissance — the rebirth of learning — was a period of flowering in the arts and in literature. It was a period when scientific discoveries were being made and when mankind's knowledge of the world was increasing. Exploration, both by land and by sea, had expanded the knowledge of the earth and had undermined fatally the existing tenets. During the later Middle Ages, there was an increasing skill in cartography, just as there was in art, and this was initially reflected in local or district maps. These small scale maps were often the result of property disputes or of the major landowners, often the church, delineating their property.

At the start of the 15th century, the production of these local maps grew dramatically. To this was added the production in 1406 of a map drawn by Ptolemy, a late Roman cartographer, of the then known world. It was drawn in a style similar to that which we would recognise today. This map was widely circulated and in 1475 the first printed version appeared. With the arrival of printed maps, the skills of the cartographer, previously limited to only a handful of people, many of them monks in the major monasteries, became more widely dispersed.

Just as the knowledge of cartography was increasing, so too were the skills associated with surveying. Although still rudimentary by modern day standards, the principle of constructing maps by actual measurement was growing in importance. Units of measurement may have varied from country to country, even district to district, but the moment that maps became scaled, so they became more useful to, for example, property owners and to mariners. The skills associated with the construction of detailed maps were also enhanced during the 16th century by the discovery of triangulation, the art by which the relative positions of places could be determined through the use of a precisely measured base line and detailed use of angles.

Many of the early cartographers of the 16th and 17th centuries were not specifically trained. Some — like Leonardo da Vinci — were artists and scientists interested in expanding human knowledge; others came from more mundane backgrounds. The great British cartographer John Speed, who flourished in the early 17th century, was a tailor by profession. A figure like Speed was able to develop as a cartographer without ever having visited the regions that he portrayed for two reasons. First, he was able to copy the work of earlier cartographers as the concept of copyright as we know it today did not exist; second, this information could only come to him through the increasing availability of detailed prints produced by craftsman, many of whom came from the Low Countries (Belgium and Holland). This latter point is of note; these craftsman-printers were producing printing plates in languages that they were not well versed in. It was inevitable that place names would be mis-spelt and these errors would be perpetuated by those using the printed maps as sources for newer publications.

By the start of the 16th century, the skills required to produce detailed scale maps were in place. Many of the earliest drawn were produced by the military — for either offensive or defensive reasons — and the military were, as the maps in this book show, to maintain an important role in cartography right through to the modern age. Many of the scale maps produced still retain elements of the older tradition of pictorial representation. To cartographers in the Renaissance, and to those working today, the pictorial representation of buildings and other facilities helps to codify. The pictorial representations that are visible in many of the maps included in this selection have three effective roles: to decorate; to provide a useful symbol (for a church or house, for example); and, to form a foundation for other information. As the draughtsman's art became more scientific, so the quality of maps improved as can be seen dramatically by material in this book.

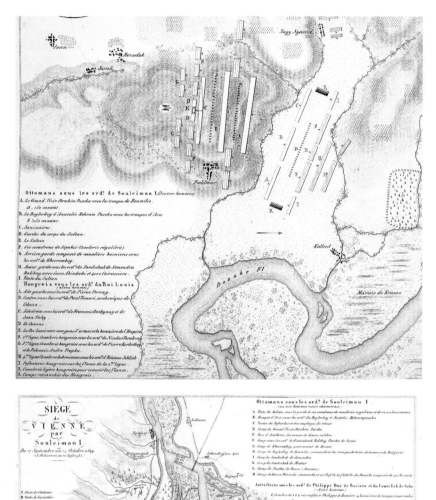

Battle of Mohács 1526

Left: Between 1520 and 1526 Muslim power expanded dramatically when the Ottoman Turks, under Suleiman I, advanced west into Hungary securing a great victory on the Danube. Still remembered as the black day of Hungarian history, the death of the Jagiellon king of Bohemia and Hungary, Louis Jagielloncyzk, during the battle was a key moment in European history. He died heirless and his possessions reverted to the Habsburgs who would control central Europe for nearly 400 years. This engraving by U. Muschani was used originally in an atlas of battles and sieges before being reproduced in J. J. Hellert's *Nouvel Atlas physique, politique et historique de l'empire Ottoman*, published by Bellizard, Dufour & Co, Paris, in 1843. No scale is shown, but there is a reference table to the distribution of Ottoman and Hungarian forces that took part in the battle.

Siege of Vienna 1529

Below Left: After their victory at the Battle of Mohács, the Turks easily overran Hungary and reached Vienna, besieging the city, unsuccessfully, in 1529. It would not be the Ottomans last attempt to take Vienna: besieged from July to September 1683, it would be saved then by Polish king Jan Sobieski. After the 1529 siege the truce of 1533 was only obtained at the price of the partition of Hungary. Western Hungary was left to its new Habsburg rulers; central Hungary, including Budapest, became an Ottoman province; Transylvania became a separate principality subject to Ottoman tutelage. This is another Muschani engraving reproduced in J. J. Hellert's *Nouvel Atlas physique, politique et historique de l'empire Ottoman* (1843). The reference tables examine the fortifications and distribution of Ottoman and Austrian forces during the siege.

Ireland 1567

Right: This is one of the oldest maps of Ireland held by the Public Records Office that can be precisely dated. The legend — in Latin — reads *Hibernia: Insula non procul ab Anglia vulgare Hirlandi vocata*, which can be translated as 'Hibernia: an island not far from England, in the common tongue called Ireland'. The map was drawn by John Goghe with annotations by Sir William Cecil (later Lord Burghley and one of Queen Elizabeth's most influential ministers). The map, which shows the whole island of Ireland and its relationship with England, Scotland and Wales, is aligned with the east-west axis forming the vertical. Pictograms delineate the various towns and cities, along with woodland and hills. With its identification of family names, it is contemporaneous with the adoption by the English authorities of a policy of colonization. This process, the Plantation, sought to transfer land ownership from the Irish native population to English settlers in the belief that the arrival of large numbers of colonists would increase the loyalty of Ireland to the English crown. It was a means by which, in theory, the power of the major Irish landowners would be reduced.

Westminster 1585

Above: Produced in Elizabeth I's reign (1558–1603) this map of the Parishes of St. Giles and St. Martin shows the rural nature of today's Soho. Outside the city of London in what would become the City of Westminster, this map shows the area only a few short decades before building started in earnest the early 17th century, and just after the rebuilding of the church of St. Martin-in-the-Fields in 1543–44. In 1536 Henry VIII had acquired St. Giles's Field as a royal hunting park for Whitehall Palace — the ancient huntsmen's cry of 'So-ho!' gave the area its name.

Cadiz Harbour 1587

Right: The wars of religion that followed the Reformation were notable for their intensity. England, protected by its position, did not suffer as many did on the Continent. Nevertheless, it was a close run thing in the late 16th century and only the might of Queen Elizabeth I's seadogs kept the fleet of Phillip II of Spain at bay. Knowing that a mighty armada was preparing for invasion, Elizabeth permitted Francis Drake to make a spoiling attack on the Spanish as the armada assembled in Cadiz in 1587. Drake scored a great propaganda victory in this raid but the next year, 1588, the Armada set sail with the intention of conquering England. This map, signed by William Borough and endorsed '1587', shows positions of English and Spanish warships and anchorages when Francis Drake 'singed the King of Spain's beard'.

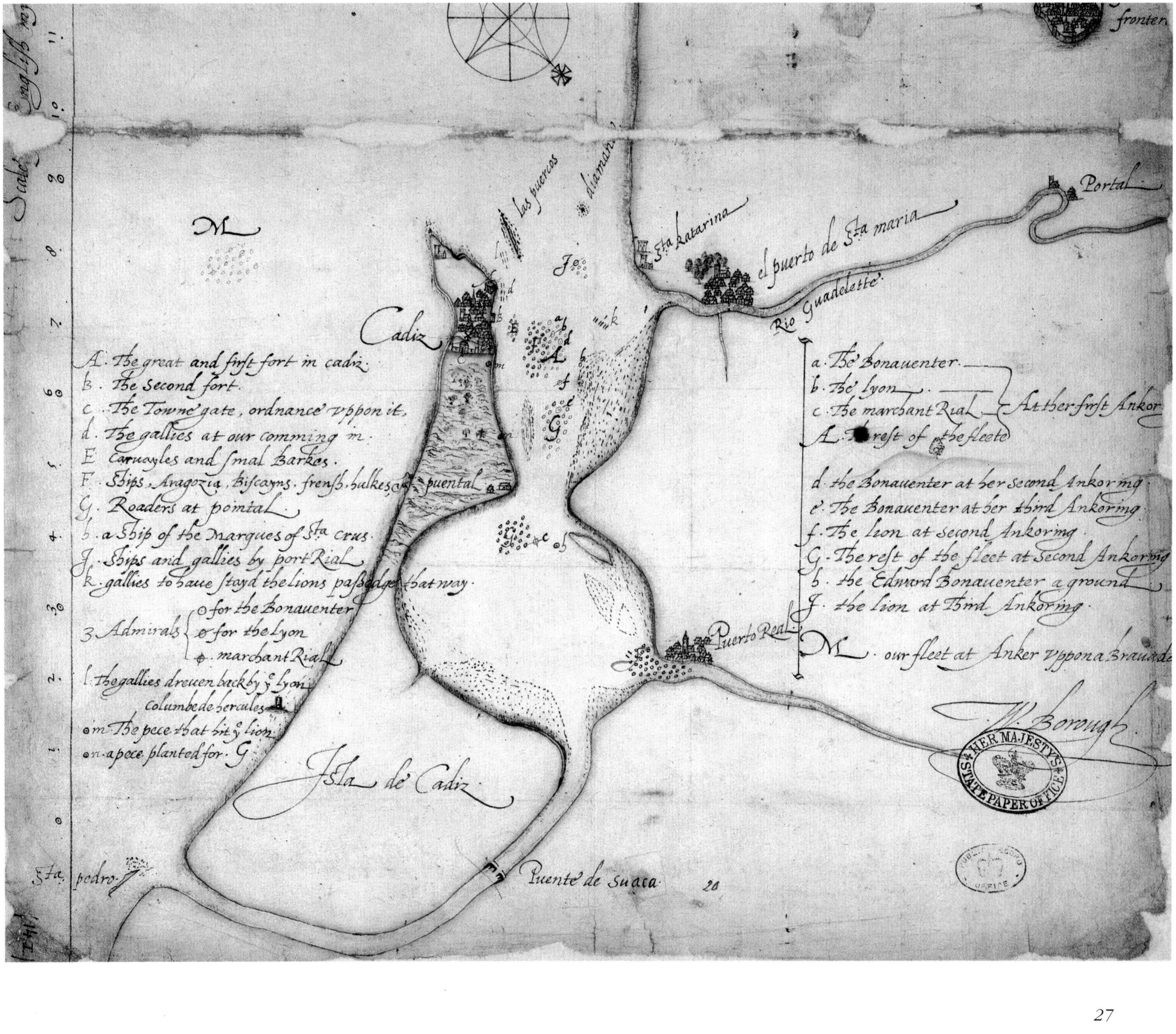

fronter

Portal

Las guercas · diamant

Sta Katarina

el puerto de Sta maria

Rio Guadelette

M

Cadiz

A. the great and first fort in cadiz
B. the second fort
c. The Towne gate, ordnance vppon it
d. the gallies at our comming in
E. Caruayles and smal Barkes
F. Ships, Aragozia, Biscayns, frensh, bulkes puental
G. Roaders at pointal
h. a Ship of the Marques of Sta Crus
J. Ships and gallies by port Rial
k. gallies to haue stoyd the Lions passadge that way

        ⎧ o for the Bonauenter
3 Admirals ⎨ o for the Lyon
        ⎩ o marchant Rial

l. the gallies dreuen back by ye Lyon
        Columbe de hercules
m. The pece that hit ye Lion
n. a pece planted for G

a. The Bonauenter
b. The Lyon
c. The marchant Rial ⎤ At ther first Ankor
A. the rest of the fleete

d. the Bonauenter at her second Ankoring
e. The Bonauenter at her third Ankoring
f. The lion at second Ankoring
G. The rest of the fleet at Second Ankoring
h. the Edward Bonauenter a ground
J. the lion at third Ankoring

M. our fleet at Anker vppon a Brauade

Puerto Real

Isla de Cadiz

Sta pedro

Puente de Suaca   20

W Borough

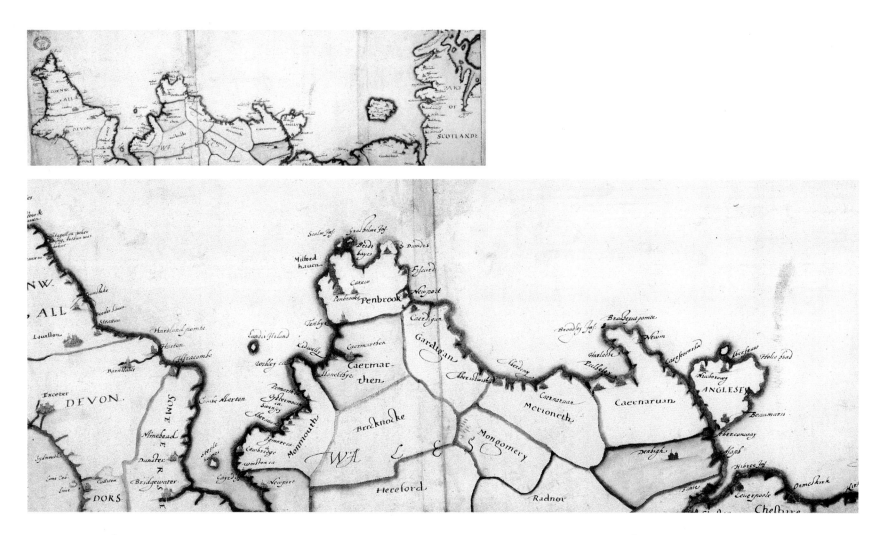

Wales 1610

Above: This map shows the coast of Britain from the Mull of Kintyre to Bridport in Dorset, concentrating on Wales. An enclosure to a larger collection of maps of Ireland, it shows the counties and main towns of Wales in the reign of James I (1603–25). It was a reign that signalled the end of a great period for Wales that had seen the Welsh house of Tudor gain the English throne. Having defeated Yorkist Richard III at Bosworth Field in 1485, Henry Tudor reigned 1485–1509 as Henry VII and his son Henry VIII (reigned 1509–47) united the two countries in two main acts of Parliament, 1536 and 1543.

Western France 1636

Right: This circular map of western France is painted on parchment with Niort at its centre. It shows an area mainly in the present départements of Vendée, Deux Sevres, Charente and Charente Maritime, and is coloured to show divisions of the country into 'élections'. The map also shows the Sévre, Charente and other rivers along with forests, principal roads and bridges. The villages and towns, including Niort and La Rochelle, are represented by perspective drawings of churches or the buildings surrounding them. The concentric circles each represent a distance of a league from Niort. It was painted by F. Granier.

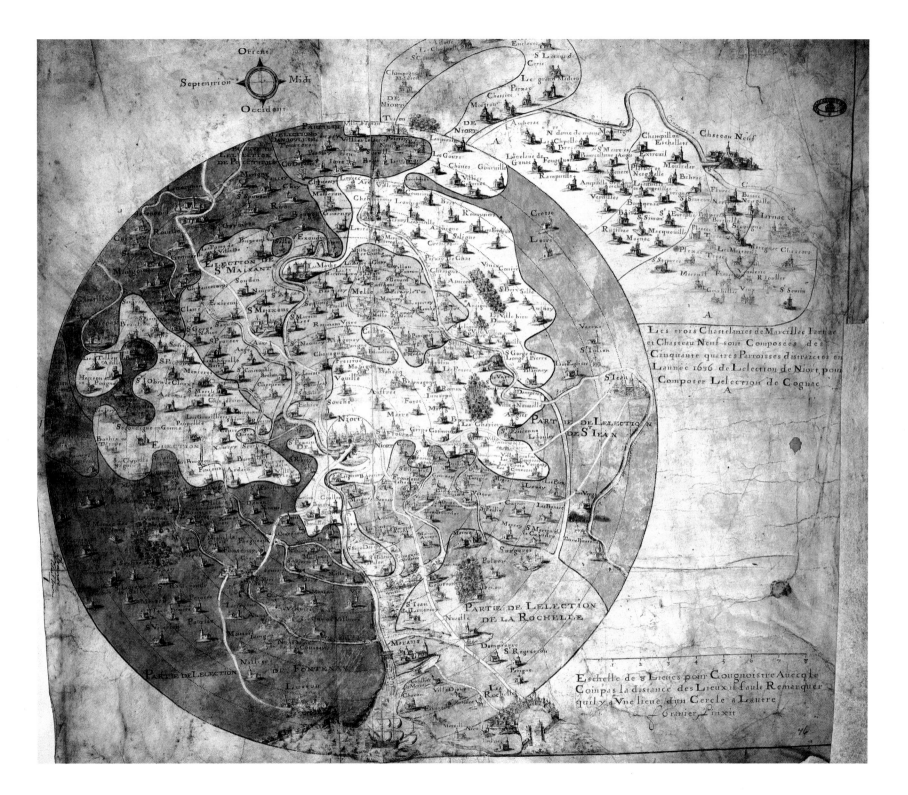

29

Portugal 1653

Left: Extracted from the Board of Trade entry book: papers relating to Portugal, 1676–79, this engraved, coloured map shows provincial and other boundaries clearly defined in colour. Portugal had been absorbed into Spain in 1580 with promises of respect for her liberties from Philip II but the provincialism inherent in the formally federal structure of the Spanish state eventually led to revolt in Portugal. By 1640 she was a state in her own right again and this map was engraved soon afterwards.

Dutch Coast 1673

Above, Left and Right: The history of England and Holland (then the United Provinces) in the late 17th century belies how close the countries would become when William of Orange ascended the English throne in 1688. There were four Anglo–Dutch naval wars between 1652 and 1684, caused mainly by maritime commercial rivalries. This coloured chart is of the coast between Nieuport and Walcheren showing the movements of the English fleet, named shoals and channels, leading lines, roads and place names on shore. The chart forms part of a letter dated 7 June 1673, from John Scott at Bruges to Sir Joseph Williamson, and contains information about Dutch preparations against the English. In June 1673, during the Third Anglo–Dutch War, the Dutch fleet defeated the English for the second time at Schooneveld (having already done so in May of the same year).

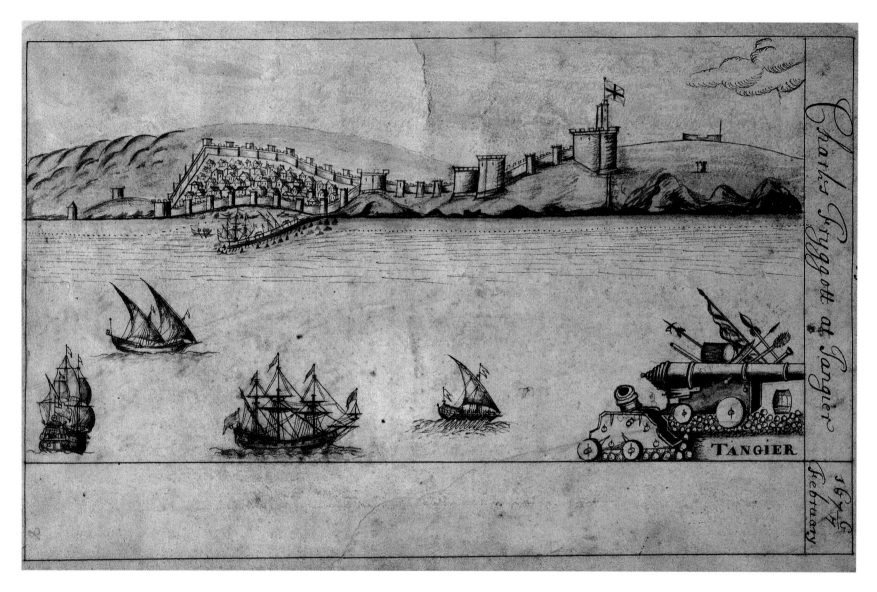

Tangier 1679

Above: This illustration was taken from the journal of Grenvil Collins, Master of H.M. Rowing Frigate *Charles*, as part of a chart of the Moroccan coast from Cartagena to Cape St. Martin showing dangerous rocks and anchorages. English fleets had first entered the Mediterranean in force under Cromwell, to pursue royalist vessels and to chastise the Barbary pirates: they would be significant players in the Mediterranean thereafter. Tangier came into English hands in 1662, as part of the dowry of Portuguese Catherine of Braganza when she married Charles II (she also brought England Bombay); it would be abandoned to pirates in 1684.

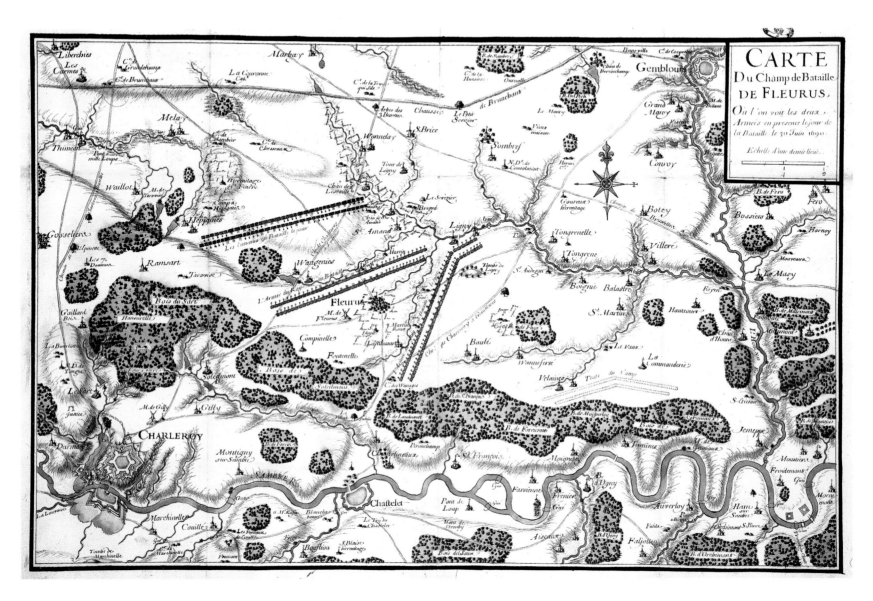

Battle of Fleurus 1690

Above: There were three battles of Fleurs, or Fleurus as it was represented. The first, on 29 August 1622, during the Thirty Years' War, saw the Spanish general Spinola defeat the Palatinate troops of the Count von Mansfeldt; the second, on 1 July 1690, during the war of the Grand Alliance, saw the French defeat the Germans and Dutch; the third, on 16 June 1794, during the French Revolutionary Wars saw the French victorious again when they defeated the Austrian army of the Duke of Coburg. This map shows the positions of the combatants in the second of these battles.

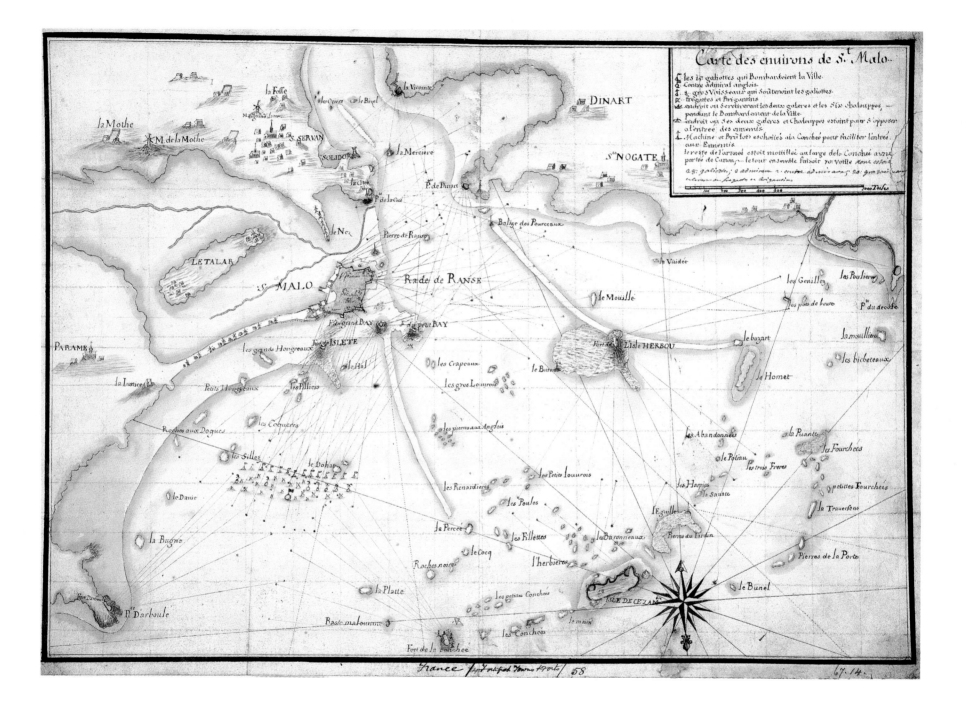

Carte des environs de St Malo.

34

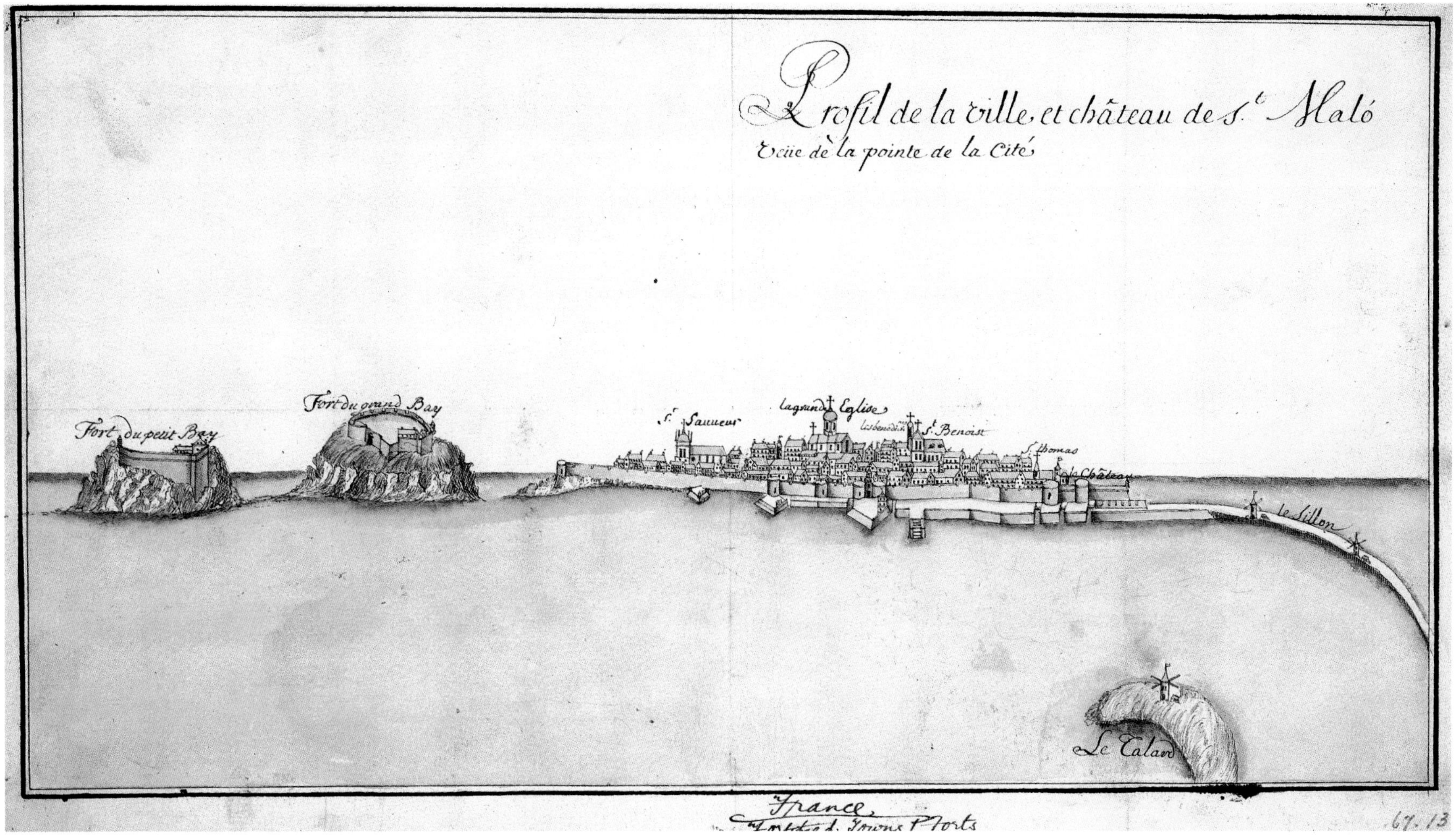

Profil de la ville, et château de S.t Malo
Vûe de la pointe de la Cité

Fort du petit Bay
Fort du grand Bay
S.t Sauueur
la grand Eglise
S.t Benoist
S.t thomas
le Château
le Sillon
Le Talard

France

St. Malo 1693

Left: The English bombarded the fortified town of St. Malo on the coast of Brittany in both 1693 and 1758. The information on this chart corresponds closely to a description of the 1693 bombardment under Admiral Benbow suggesting this was the chart of that action. It shows the bombardment, the French batteries with their ranges and positions of French ships.

St. Malo 1700

Above: An early 18th century coloured drawing of St. Malo on the Rance estuary. As shown by the illustration at left, the town was far from peaceful at this time. It was the haven of such famous corsairs as Duguay Trouin (1673–1736) and Surcouf (1773–1826) who inflicted significant losses on Dutch, Spanish and English shipping. Indeed, in the 17th and 18th centuries, these sailors received special license to pirate allowing them to attack ships without fear of punishment. In St. Malo, nearly as many ships were fitted out for piracy as for legitimate trade with the East Indian Company. The Fort National (formerly called the Fort Royal) dates back to 1689 and was built by Louis XIV's famous military architect Vauban. In the 18th century the city walls were extended and modified.

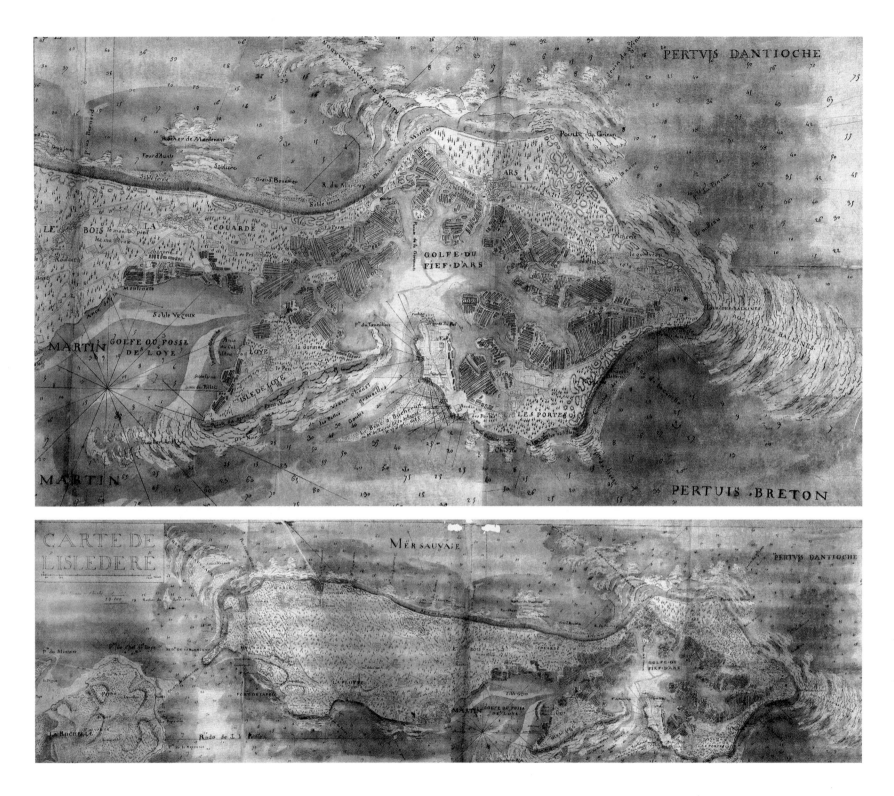

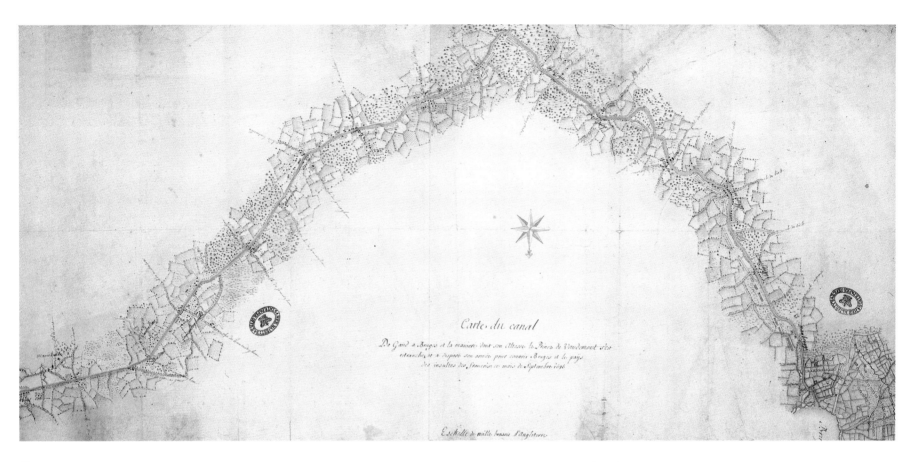

Carte du canal

De Gand a Bruges et la maniere dont son Altesse le Prince de Vaudemont s'est retranché, et a disposé son armée pour couvrir Bruges et le pays des insultes des François ce mois de Septembre 1696.

Eschelle de mille brasses d'Angleterre.

Île De Ré 1700

Left: This early 18th century coloured map on tracing paper, shows the Île de Ré, an island just off the French coast near La Rochelle. Famous for its early lighthouse, it was of great strategic importance and in St. Martin boasted a strong fortified major town — the English had attacked the island in 1627 as part of its attempts to help the Huguenots in La Rochelle. Today, the island, an up-market tourist resort, is joined to the mainland by the Sablanceau bridge; at the point where the cliffs of Sablanceau are identified on this map.

Ghent Canal 1696

Above: 'Map of the canal from Ghent to Bruges and the way in which his Highness the Prince of Vaudemont repositioned his forces to cover Bruges and its environs against incursions from the French'. Flanders suffered from almost continuous conflict in the 17th and early 18th century, frequently at the hands of the French. During the period Ghent, noted for its cloth trade, searched for a means to improve its trading position and began construction of a canal between the city and Ostende, passing through Bruges. It was only after the War of the Spanish Succession and the more enlightened rule of the Austrian Habsburgs that prosperity returned.

THÉATRE DE LA GUERRE
DANS
LA FLANDRE ET LE BRABANT,
DE 1703 A 1712,
Dressé au Dépôt général de la Guerre,
SOUS LA DIRECTION DE Mr LE Lt GÉNÉRAL PELET.
Légende des Lignes défensives.
H Lignes de la Mer à la Meuse par Nieuport, Commines, la Lys, la Deule, la Scarpe, la Haine, la Sambre jusqu'à Namur et par la Meuse jusqu'à Mezieres. - 1706-7.
I Ligne de Quinchy à Bonfontaine. 1709.
K Ligne de Valenciennes à la Forêt de Mormal par le Quesnoy. - 1709.

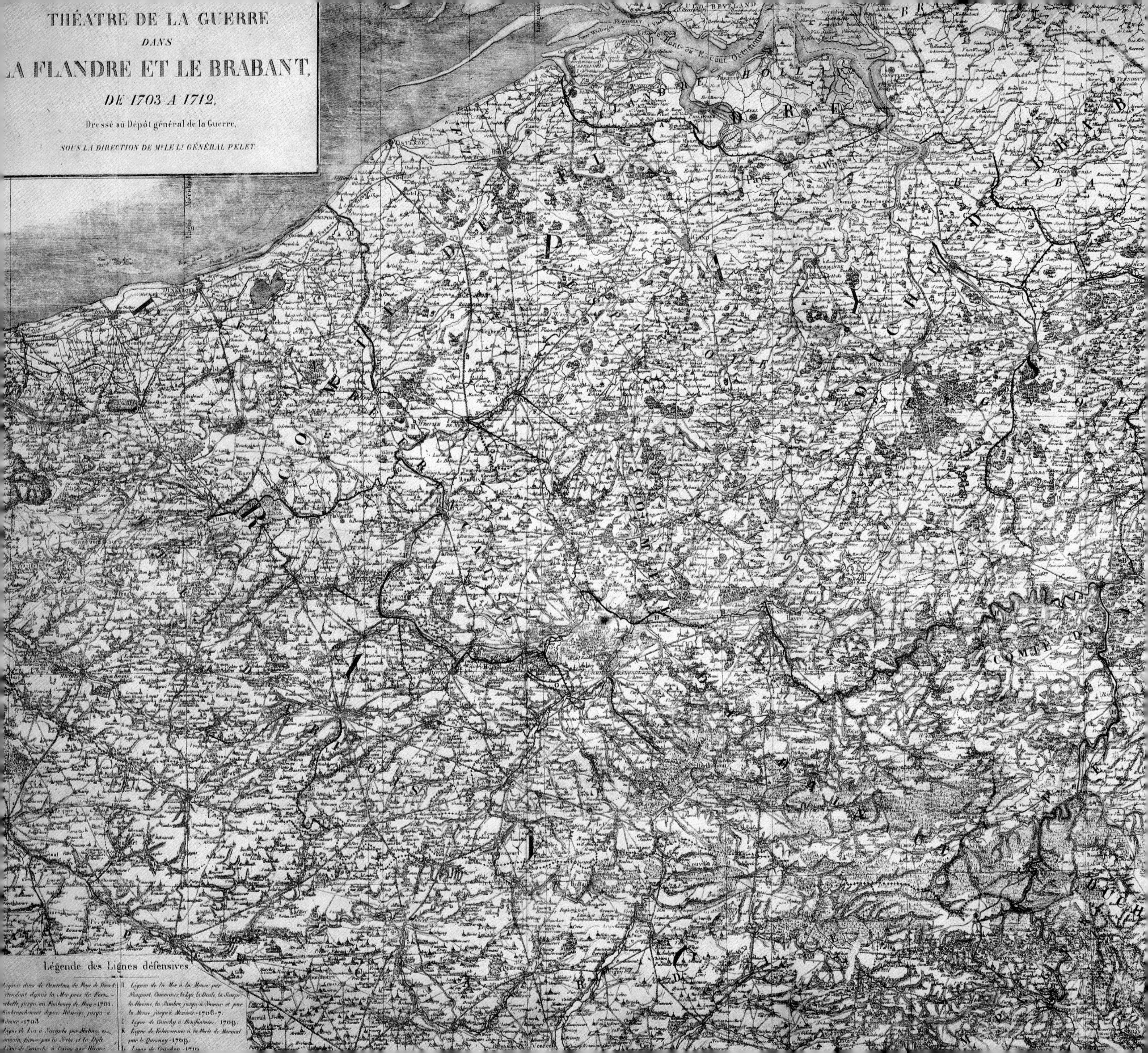

Belgium and the Netherlands 1701–12

Left: Compiled between 1830 and 1845 in Paris under the direction of Lieutenant-General Pelet, director of the Dépôt général de la Guerre, this map shows the theatre of war in Flanders and Brabant that saw so much action in the early part of the 18th century. The culmination of Louis XIV's wars of aggrandisement, the War of the Spanish Succession began in 1701 when the Grand Alliance of The Hague was formed by England, the United Provinces and the Holy Roman Empire against France and Spain. It purported to be about who would rule in Spain following the death of the last Habsburg monarch, Charles II. However, Louis XIV identified its main import: control of trade to the New World. In 1702 the Duke of Marlborough became Captain-General of the English forces: he would defeat the French at Blenheim (see below), Ramillies in 1706 and in 1708, alongside Prince Eugene of Austria, at Oudenarde.

Blenheim 1704

Right: This engraved map represents a plan of the Duke of Marlborough's famous victory on 13 August 1704 at the Bavarian village of Blenheim over Marshal Tallard in command of the Franco–Bavarian forces. The victory was one of the decisive moments in the War of Spanish Succession, stopping the French threat to Vienna and the central Habsburg Empire. Marlborough — whose portrait is in the upper left corner — accomplished this feat with a truly international army of Dutch, Germans, Danes and British.

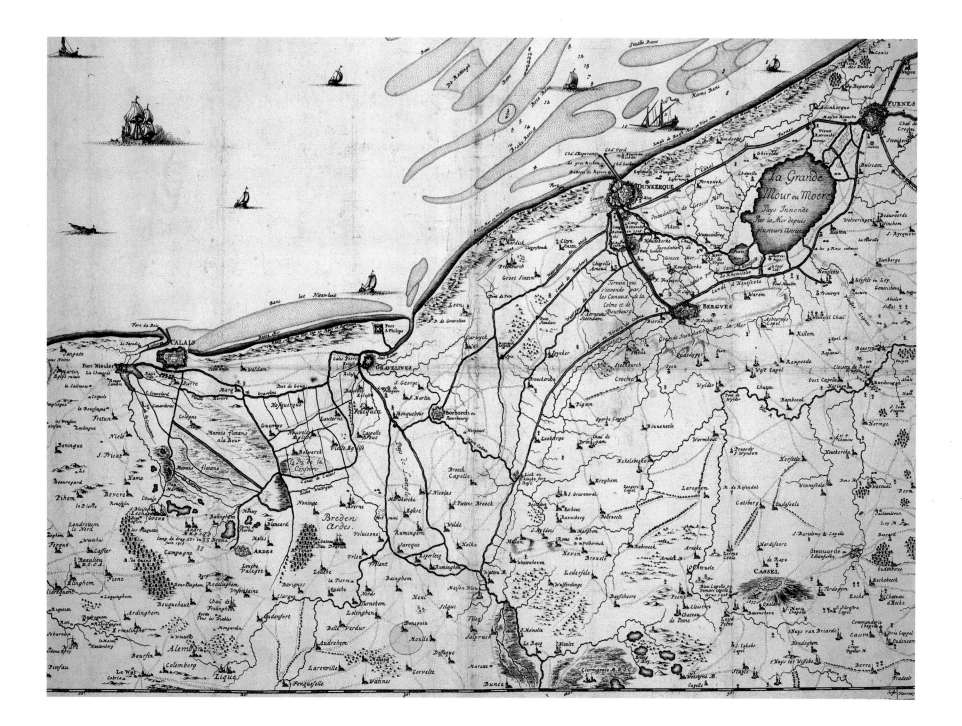

Pas de Calais 1707

Left: This engraved map shows the 'Environs of Dunkerque, Bergues, Furnes, Gravelines, Calais, et Autres'. This area of Northern France and Belgium has always of importance to the British War Office; the map shows rivers, sandbanks, soundings, woodland, roads, tracks, canals, shipping, towns, villages and forts.

The Rhine 1711

Right: This colourful map illustrates another action in the War of the Spanish Succession when French forces attacked the rearguard of the Austrian army near Fort Louis on the Rhine on 23 August 1711. The detailed alphabetical key identifies the units involved.

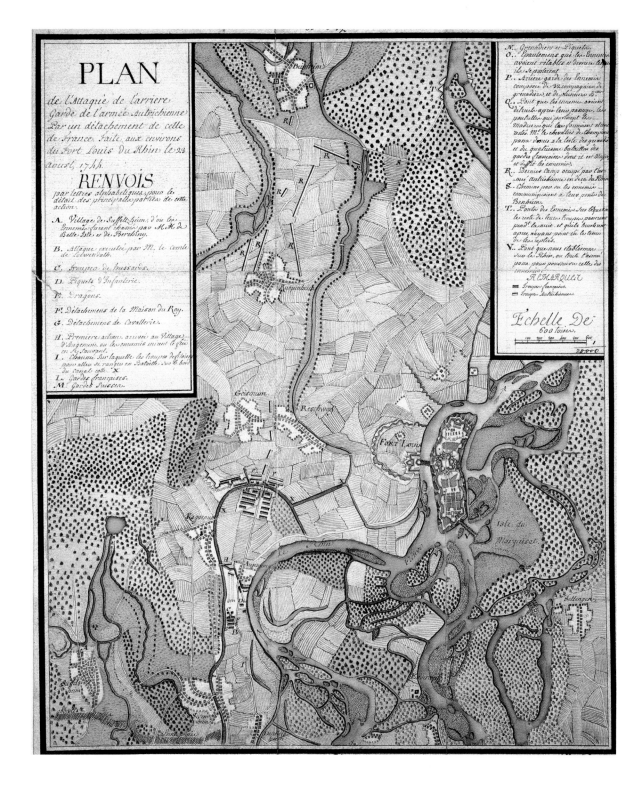

...ICE D'ALLEMAGNE, ou sont distingués ... S ESTATS DE LA M. L'AUSTRICHE, DES ESLECTEURS ECCLESIASTIQUES DE ME... SOUABE, FRANCONIE, HESSE, WESTPH...
...TIN DU RHEIN, DU DUC DE SAXE, et DU MARQUIS DE BRANDEBOURG, et de tous LES ETATS, et SOUVERAINETES qui sont dans la ...
...AM d'Austriche en Allemagne, sont le Royaume et Eslectorat de Boheme, l'Archiduche d'Austriche, les Duches de Stirie, Carinthie, Carniole, la Principauté de Souabe, le Comté ... de Tyrol &c a l'Empereur la Franche Comté, au Roy et...

MER D'ALLEMAGNE

MER

ou MER

DE NORD

B.O

ROYAUME D'ANGLETERRE

ZUYDER ZEE

WESTPHALIE

CERCLE DE WESTPHALIE

LUXEMBOURG

LORRAINE

PALATINAT DU RHEIN

LANDGRAVIAT DE HESSE

DUCHE DE CLEVES

EVECHE DE MUNSTER

COMTE DE HOIA

DUCHE DE BREME

DUCHE DE HOLSTEIN

DUCHE DE MEKLENB...

DUCHE DE SAXE

PRINCIPAUTE D'ANHALT

MAGDEBOURG

LUNEBOURG

BRUNSWIK

HILDESHEIM

COMTE DE LIPPE

RAVENSBERG

PADERBORN

EVECHE DE COLOGNE

DUCHE DE JULIERS

COMTE DE NASSAU

DILLENBOURG

HESSE DARMSTAT

FULDE

WORTZBURG

CERCLE DE FRANCONIE

NURENBERG

BAMBERG

MAYENCE

COMTE D'ARTOIS

HAYNAU

PICARDIE

ROYAUME DE FRANCE

BRABANT HOLLANDOIS

SEIGNEURIE DE FRISE

GUELDRE

COMTE DE ZUTPHEN

German Empire 1715

Left and Below: This engraved, coloured map of central Europe is by I. Covens and C. Mortier of Amsterdam in c.1715. It shows 'Germaniae l'Empire d'Allemagne . . . tous les Etats Principautés et Souveraintées qui passent . . . sous le nom d'Allemagne'. With its scale bars in Italian, German, French, Polish and Hungarian leagues and a reference table to roads, towns, villages, universities etc., it is a remarkable piece of work and shows the difficulties that the German 'Empire' caused cartographers of the period. Later maps will chart the unification of Germany. In 1715, at the signing of the final part of the Treaty of Utrecht that ended the War of the Spanish Succession, Germany as we know it today was still composed of a myriad states, many of which were part of that medieval survivor, the Habsburg-dominated Holy Roman Empire. Despite the animosity felt towards the Austrian Habsburgs in the German states — and the views of 19th century Prussian propaganda — it is wrong to overemphasise the fragmentary nature of the Holy Roman Empire. At the time it was seen by most politicians as an important element in the balance of European power.

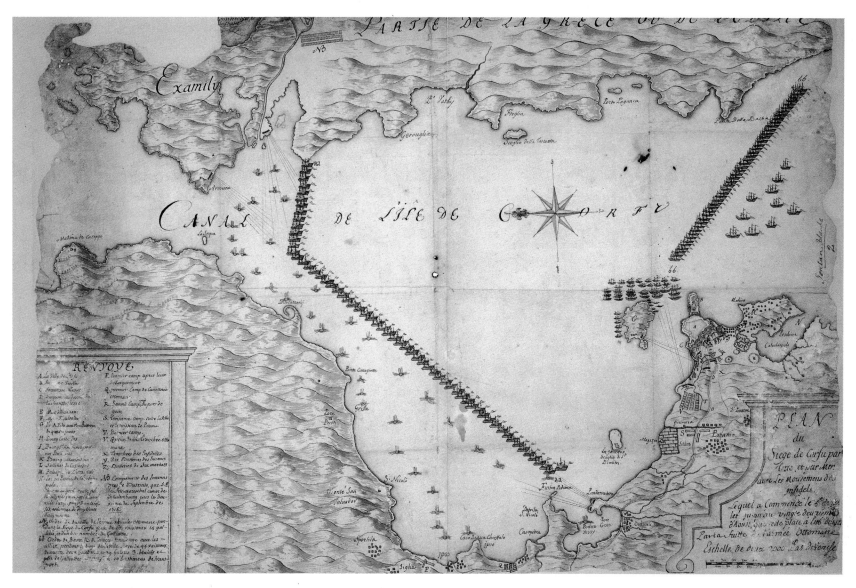

Corfu 1716

Above: For centuries the Turks struggled with the Venetians for control of the Ionian islands because of their strategic importance to the maritime trade routes through the Adriatic. For Venice the 18th century was a period of decline. The last real achievement of her armed forces was the successful defence of Corfu against the Turks in 1716; the Venetian forces were led, however, by a German general. Venice managed to cling on to Corfu and its other possessions until the demise of the republic at the hands of Napoleon in 1797 when it became part of Austria. This map shows the Turkish siege of Corfu, identifying the positions of the combatants.

Belgrade 1717

Right: This is another Muschani engraving (see also page 23) reproduced in J. J. Hellert's *Nouvel Atlas physique, politique et historique de l'empire Ottoman* (1843). It shows the plan of the battle of Belgrade, 16 August 1717, between the Ottoman Turks under the Grand-Vizir Khalil-Pascha and the Austrians under Prince Eugène. During a renewed struggle with the Habsburgs (1716–18) the Turks suffered several heavy defeats during which they lost Belgrade which had been a Turkish fortress for over two centuries. While still a potent force, the Turks were not the power they had been in the 17th century and Hungary was able to mount a successful campaign to regain lost territory.

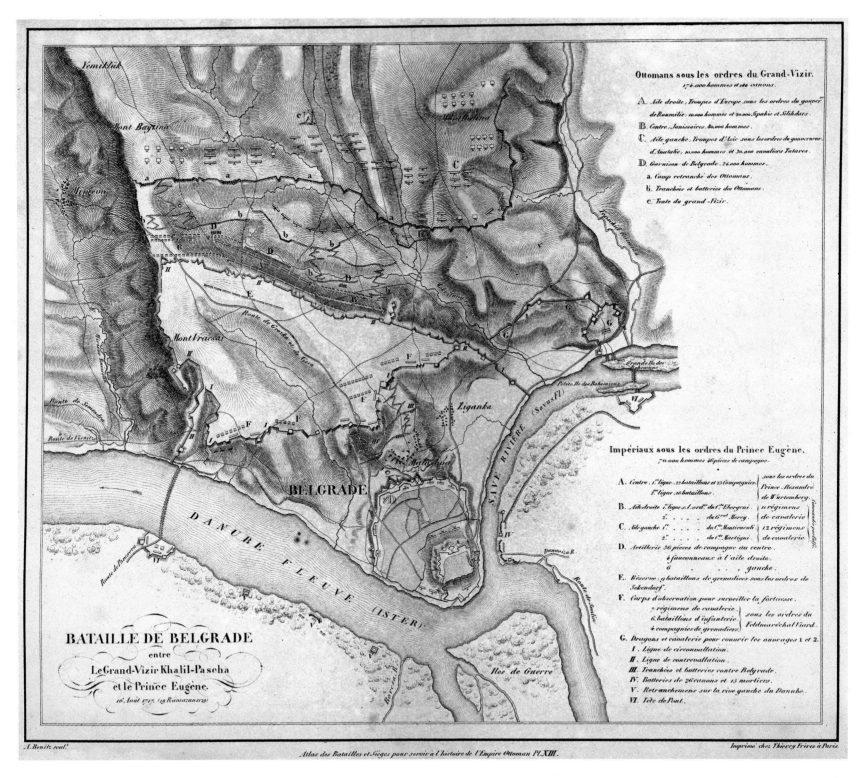

Ottomans sous les ordres du Grand-Vizir.
174.000 hommes et 180 canons.

A. *Aile droite, Troupes d'Europe sous les ordres du gouver.*
   de Roumélie, 10.000 hommes et 20.000 Spahis et Silihdars.

B. *Centre, Janissaires, 80.000 hommes.*

C. *Aile gauche, Troupes d'Asie sous les ordres du gouverneur*
   d'Anatolie, 10.000 hommes et 30.000 cavaliers Tatares.

D. *Garnison de Belgrade, 24.000 hommes.*

   a. *Camp retranché des Ottomans.*

   b. *Tranchées et batteries des Ottomans.*

   c. *Tente du grand-Vizir.*

Impériaux sous les ordres du Prince Eugène.
70.000 hommes 86 pièces de campagne.

A. *Centre, 1ère ligne, 32 bataillons et 23 compagnies.* *sous les ordres du*
   2e ligne, 18 bataillons. *Prince Alexandre de Wurtemberg.*

B. *Aile droite 1ère ligne s.l.ord. du Cte Ebergeni.* *11 régimens*
   du Gral Mercy. *de cavalerie*

C. *Aile gauche 1ère ... du Cte Monticuculi.* *12 régimens*
   2e ... du Cte Martigni. *de cavalerie*

D. *Artillerie 36 pièces de campagne au centre.*
   4 fauconneaux à l'aile droite.
   6 ... gauche.

E. *Réserve, 9 bataillons de grenadiers sous les ordres de*
   Sekendorf.

F. *Corps d'observation pour surveiller la forteresse.*
   7 régimens de cavalerie. *sous les ordres du*
   6 bataillons d'infanterie. *Feldmaréchal Viard.*
   4 compagnies de grenadiers.

G. *Dragons et cavalerie pour couvrir les ouvrages 1 et 2.*

   I. *Ligne de circonvallation.*

   II. *Ligne de contrevallation.*

   III. *Tranchées et batteries contre Belgrade.*

   IV. *Batteries de 26 canons et 13 mortiers.*

   V. *Retranchemens sur la rive gauche du Danube.*

   VI. *Tête de Pont.*

BATAILLE DE BELGRADE
entre
Le Grand-Vizir Khalil-Pascha
et le Prince Eugène.
16 Août 1717. (19 Ramazan 1129)

A. Benitz sculp.

Atlas des Batailles et Sièges pour servir à l'histoire de l'Empire Ottoman Pl. XIII.

Imprimé chez Thierry Frères à Paris.

45

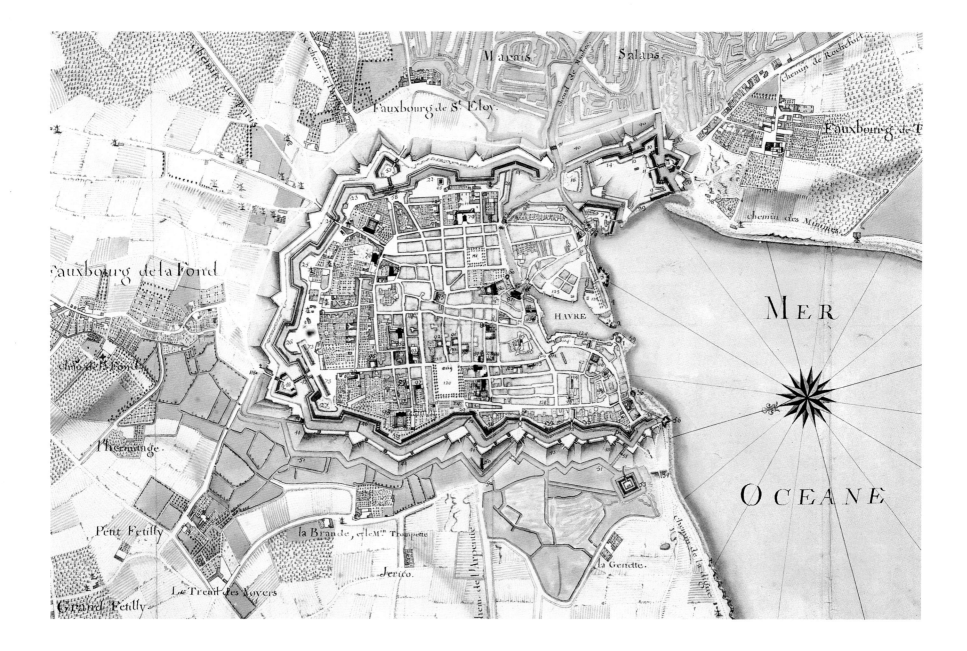

MER

OCEANE

Marais Salans

Fauxbourg de St. Eloy.

Fauxbourg de la Fond

HAVRE

l'hermitage

Petit Fetilly

Grand Fetilly

Le Treuil des Loyers

la Brande, et le Mn. Trompette

Jerico.

chemin de l'Arpentie

la Genette.

chemin de la digue

Fauxbourg de T

chemin des Minimes.

chemin de Rochefort

La Rochelle 1718

Left: A coloured map showing the town of La Rochelle, its fortifications and the surrounding countryside. In 1628, despite attempts by the English to stop him, Cardinal Richelieu — Louis XIII's chief minister — took the Protestant stronghold of La Rochelle. After this the Huguenots were never again a serious threat to the unity of France.

Aragon 1719

Right: Engraved by P. Starckman in 1719, this map of the Kingdom of Aragon is dedicated to Philippe duc d'Orleans, regent in France following the death of Louis XIV in 1715 until Louis XV's majority in 1723. Despite his reputation as a roué, Philippe was a patron of the arts, and mapmaker Bourgignon d'Anville, geographer to the king, obviously knew which side his bread was buttered! At the time this map was compiled the Kingdom of Aragon was just ending an existence of some 500 years: just as England and Scotland had been joined by the Act of Union in 1707, so Aragon, Valencia (in the same year) and Catalonia (in 1714) lost their political autonomy in the union of Spanish kingdoms.

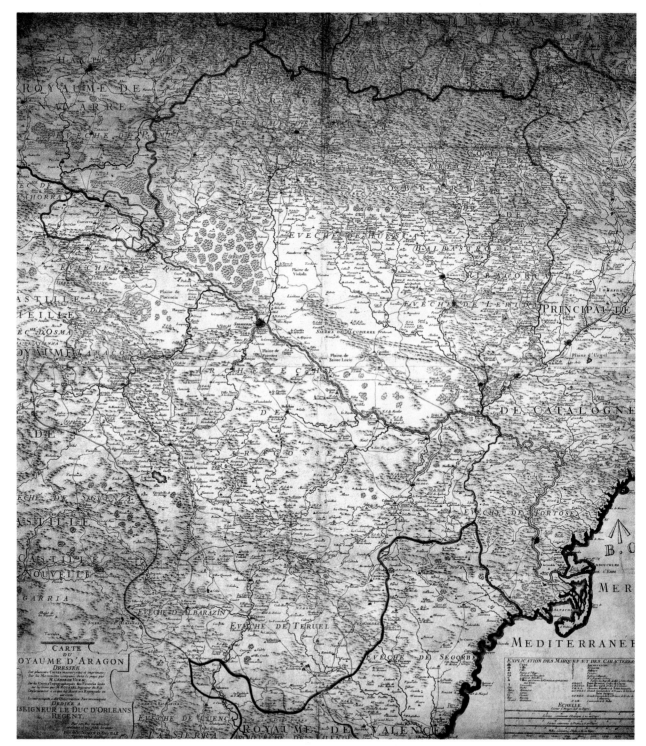

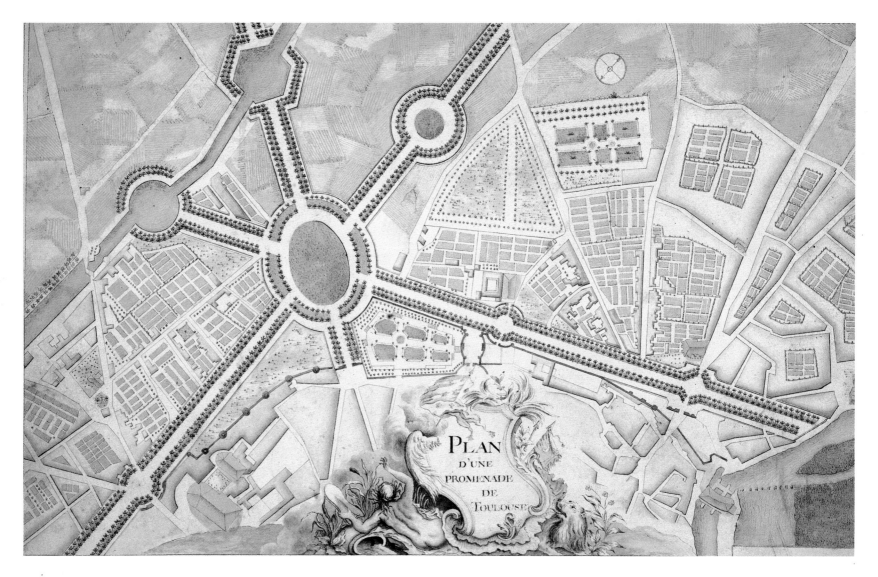

Toulouse 1720

Above: A delightful, coloured plan of Toulouse in 1720. It shows an oval with grass, trees and paths from which radiate six avenues of trees, built-up areas, cultivated areas, formal gardens, a canal and a river. Toulouse saw much building work in the 18th century, including the huge Hôtel de Ville in the heart of the city in the Place du Capitole and a ring of fine boulevards built in the 18th and 19th centuries.

Low Countries 1730

Right: Showing the River Scheldt and Anvers, this map of 1730 came from the papers of William Pitt the Younger (1759–1806), British first minister 1783–1801 and 1804–06. Following the War of the Spanish Succession, the Spanish Netherlands, which had been governed by the Elector of Bavaria, became part of the Austrian Habsburg Empire under the terms of the Treaties of Rastadt and Baden in 1714. For some time afterwards, the Scheldt was closed to commercial traffic by the United Provinces.

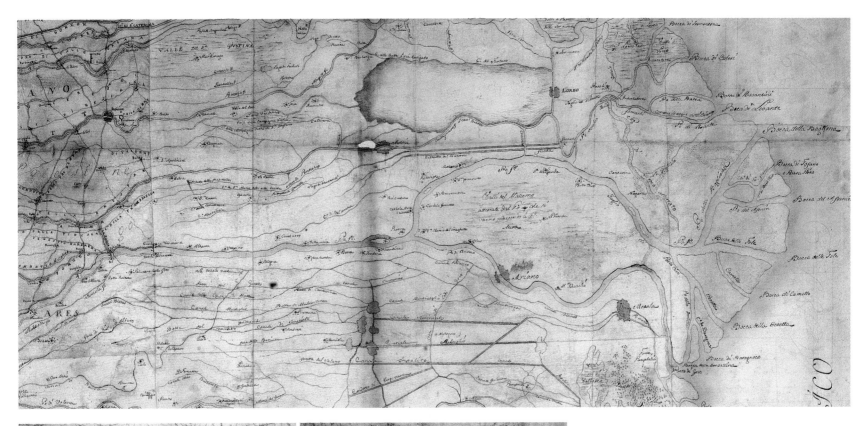

Eastern Lombardy 1736

Above, Details left and far left: Map of Lombardy from Mantua (top left) to the Adriatic Sea. The River Po runs along the top part of the map; the foothills of the Appennines encroach on the bottom left. Centred on Milan, Lombardy had long been a part of the Spanish Empire. Acquired in 1713 by the Austrian Habsburgs, it would prosper in the 18th century thanks to the fertility of the region fed by the waters of the Po, the strong and wealthy towns of the area and the enlightened rule of the Habsburgs. At bottom right of the map, on the Adriatic Coast, is Ravenna in Emilia Romagna.

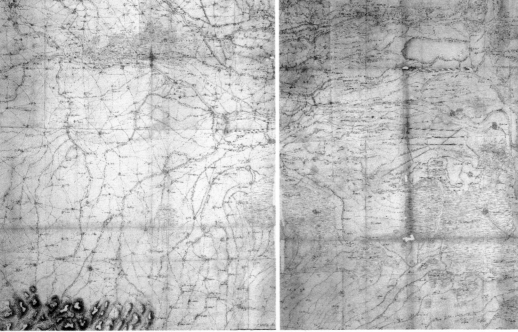

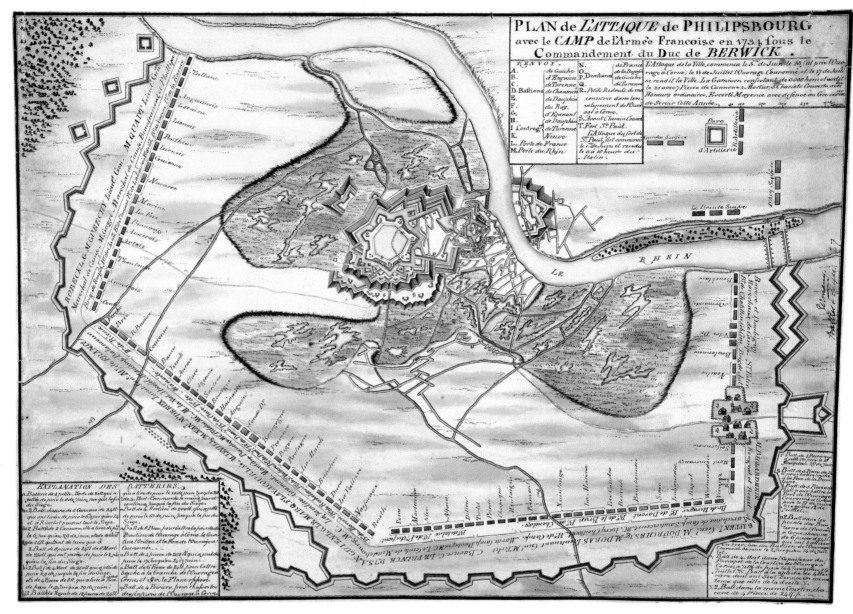

PLAN de L'ATTAQUE de PHILIPSBOURG avec le CAMP de l'Armée Françoise en 1734 sous le Commandement du Duc de BERWICK.

Siege of Philipsbourg 1734

Above: The War of the Polish Succession saw the Russians and Habsburgs proposing Augustus II of Saxony against the French candidate, Stanislas Leszcynski. It took place between 1733 and 1735 and much of the action centred on Italy and the Rhineland, where the siege of Philipsbourg, illustrated here, took place between 1 June and 17 July 1734. The detailed keys identify the artillery involved in the siege.

Ukraine 1736

Right: Engraved chart produced 'par ordre de l'Imperatrice de toutes les Russies' — the empress being Anna (1693–1740). During her reign Russian armies made progress both in the west, during the War of Polish Succession, and in the south where they regained Azov in the Russo–Turkish War of 1735–39. This map shows Russian military operations in the area; Azov (Avoff) is at right.

CARTE de la PETITE TARTARIE Dressée par ordre de l'Imperatrice de toutes les Russies

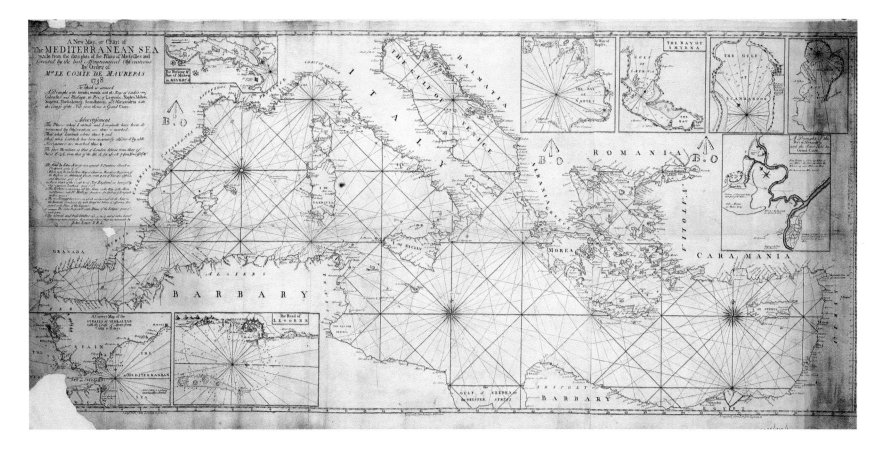

Mediterranean Sea 1738

Above: 'A New Map, or Chart of the Mediterranean Sea made from the draughts of the Pilots of Marseilles and Corrected by the best Astronomical Observations By Order of M. le comte de Maurepas 1738'. This map of the Mediterranean is given added value by the inclusion of detailed insets such as 'A Correct Map of the Straits of Gibraltar with the coast of Spain from Cadis to Malaga'. It was sold by John Senex of Fleet Street, 'over against St. Dunstan's church 2s 6d', who advertises his range of publications at left.

Prague 1742

Right: Having suffered the Wars of the Spanish and Polish successions, a third 18th century succession dispute led to conflagration in the form of the War of the Austrian Succession. On one side were Britain and the United Provinces in support of Maria Theresa of Austria; on the other the aggressive Frederick II of Prussia and France supporting Charles Albert of Bavaria. During the course of the war, which lasted from 1740 to 1748, King George II became the last British monarch to enter battle, at Dettingen in 1743. This map shows the two sides, France and 'l'ennemies' — Austria — ready for battle around Prague in 1742.

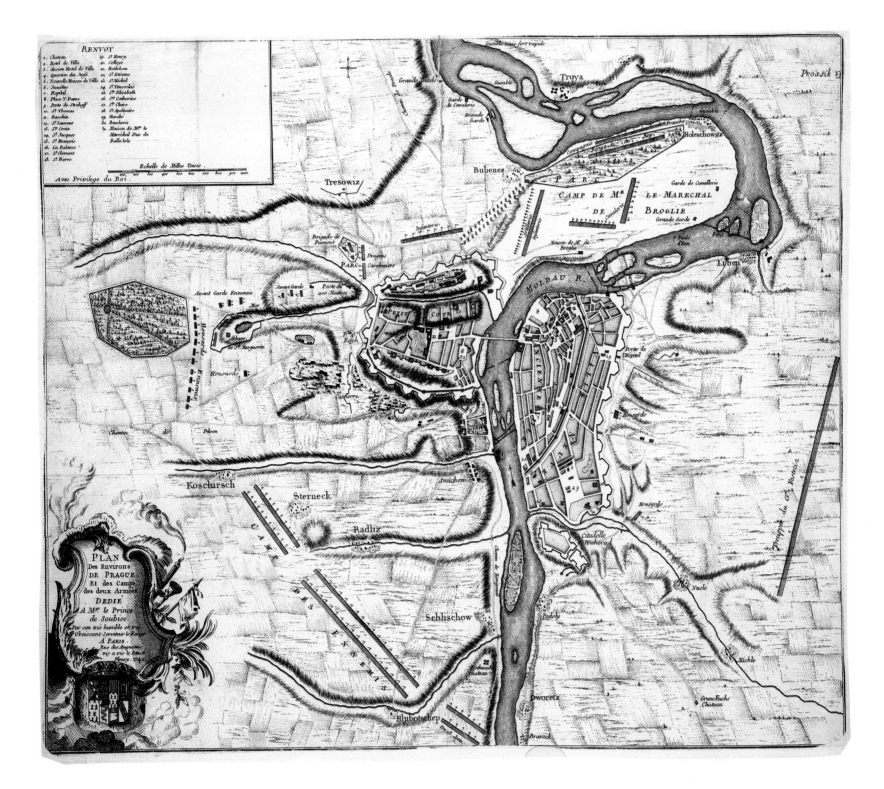

RENVOY

1. Chateau
2. Hotel de Ville
3. Ancien Hotel de Ville
4. Quartier des Juifs
5. Nouvelle Maison de Ville
6. Jesuittes
7. Hopital
8. Place N. Dame
9. Porte de S. Wohoff
10. St Thomas
11. Rarchin
12. St Laurent
13. St Croix
14. St Jacques
15. St François
16. La Balance
17. St Clement
18. St Pierre
19. St Henry
20. Collège
21. Betlehem
22. St Etienne
23. St Bickel
24. St Vencesles
25. St Elizabeth
26. St Cathrine
27. St Claro
28. St Apolinaire
29. St Apolinaire
30. Marche
31. Boucherie
32. Maison de Mgr le Marechal Duc de Belle Isle

Echelle de Milles Toises

Avec Privilege du Roi.

PLAN
Des Environs
DE PRAGUE
Et des Camps
des deux Armées
DEDIÉ
A Mgr le Prince
de Soubise
Par son très humble et très
Obeissant Serviteur le Rouge
A PARIS
Rue des Augustins
vis a vis le Panier
Fleury, 1742.

Tresowiz

Troya

Proissik

Holeschowitz

Bubenez

PARC

CAMP DE Mr LE MARECHAL
DE BROGLIE

Grande Garde

Garde de Cavalerie

Infanterie

Maison de Mr de Broglie

Luben

Brigade de Piemont

PARC

Dragons Caraviniers

MOLDAU R.

Avant Garde Ennemie

Avant Garde

Porte de 200 Mathor

PETIT COTÉ

St Roch

Porte de l'Hopital

Howard de l'Ennemie

Ile de S. Marguerite

VOLVNE

Koschrsch

Smichow

Sterneck

Radliz

Citadelle Wichocrad

Nusle

Houzauxle

CANAL DES ENNEMIS

Schlischow

Padoly

Dworetz

Michle

Hlubotschep

Branick

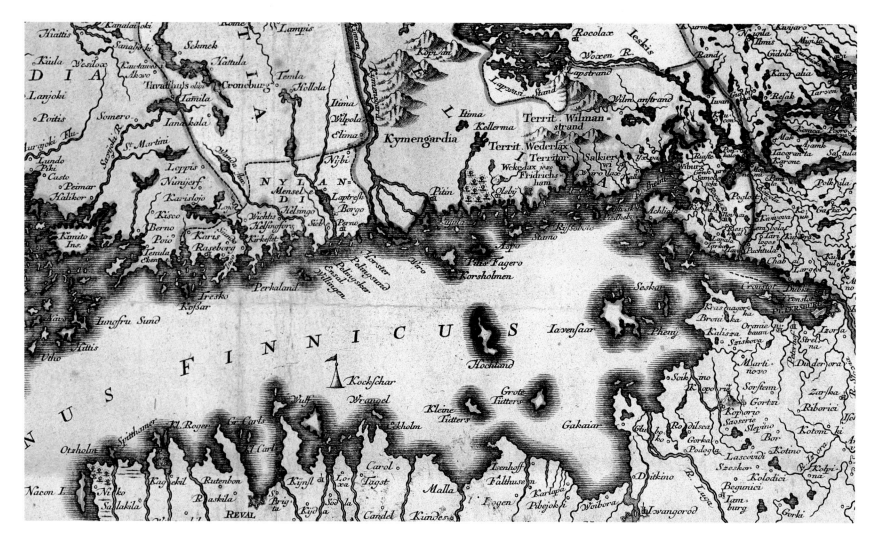

Finland 1743

Right and Detail Above: The Foreign Office index for this map gives a date of around 1720 — certainly a logical date given that the Great Northern War ended at about that time and Finland, taken by the Russians in 1713–14, was handed back to Sweden by the Treaty of Nystad (1721) — except for Kexholm, visible in red at the right of the image above. Unfortunately the name on the map is Tobiae Conradi Lotter, who wasn't born until 1717. It is likely, therefore, that rather than being dated 1720, the map was engraved following the second return of Finland to Sweden at the Peace of Abo following the war of 1741–43.

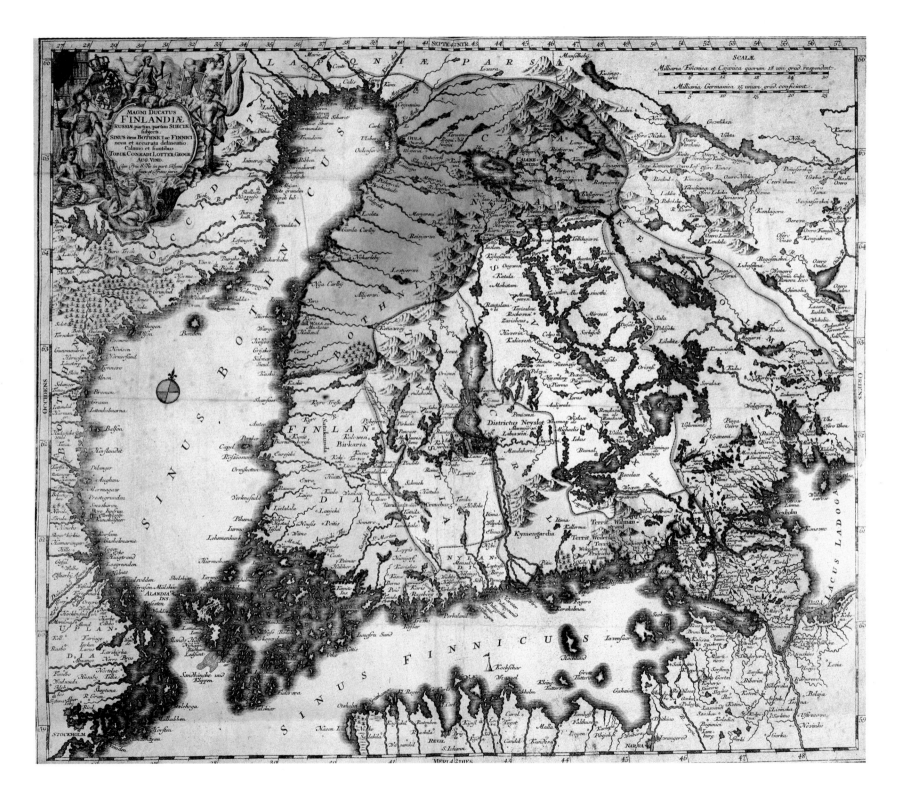

57

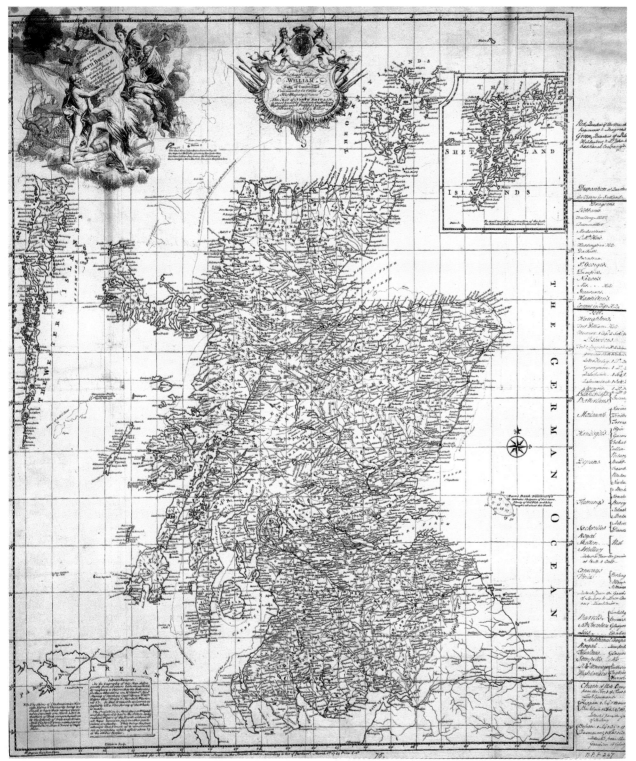

Scotland 1745/6

Left: 'A New and Correct Mercator's Map of North Britain carefully laid down from the latest surveys and most approved observations by Elphinstone, Engineer, 1746' dedicated to 'His Royal Highness William Duke of Cumberland Commander-in-Chief of His Majesty's Forces' — better known as 'Butcher' Cumberland, who commanded the king's forces against the Jacobites at Culloden on 16 April 1746, and in the reprisals that followed. Although the draughtsman calls it 'North Britain', this map of Scotland shows the disposition of quarters for British troops in the country and Wade's roads allowing them speedy movement if needed. Following the unpopular Act of Union in 1707, there were many planned Scottish risings linked with the Jacobite cause, most often aided and abetted by the French. Only two actually took place: that of 1715 by 'James III' of England — the Old Pretender, son of deposed King James II; and that of 1745, when the Old Pretender's son, 'Bonnie Prince Charlie' reached Derby before being hounded back up country and fleeing to France, with his army defeated at Culloden.

Brussels 1745

Right: This engraved map of Brussels is dated 1745 and has detailed reference tables identifying the main features — and one or two pencilled updates. Brussels takes its name from Broekzele, meaning 'Village of the Marsh' and is situated in the wide Senne River floodplain.

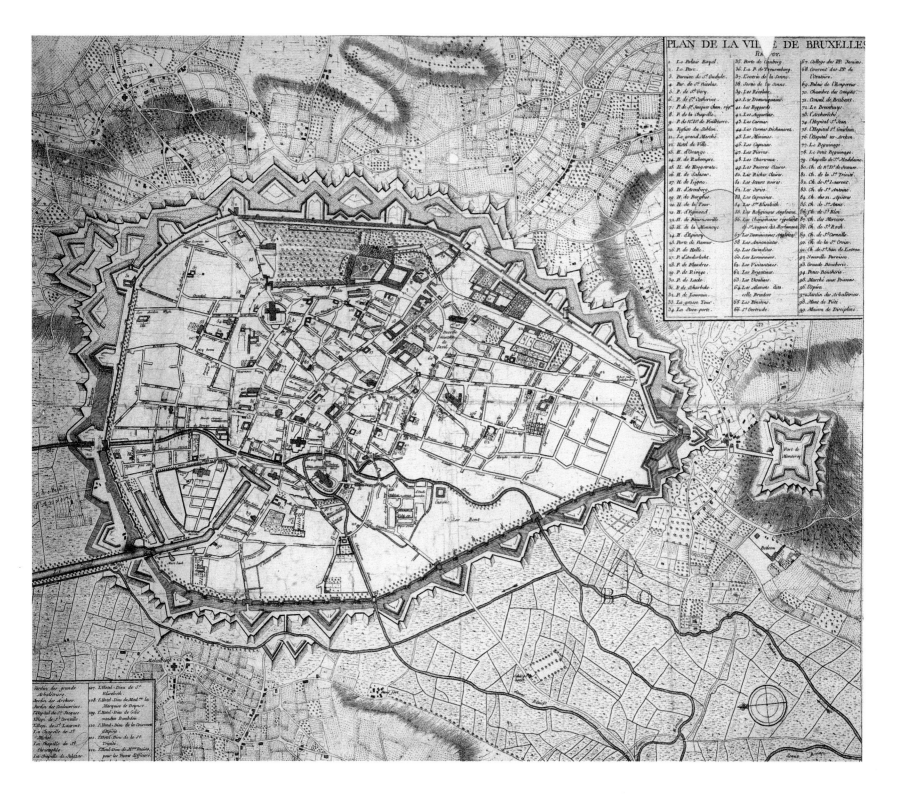

PLAN DE LA VILLE DE BRUXELLES

59

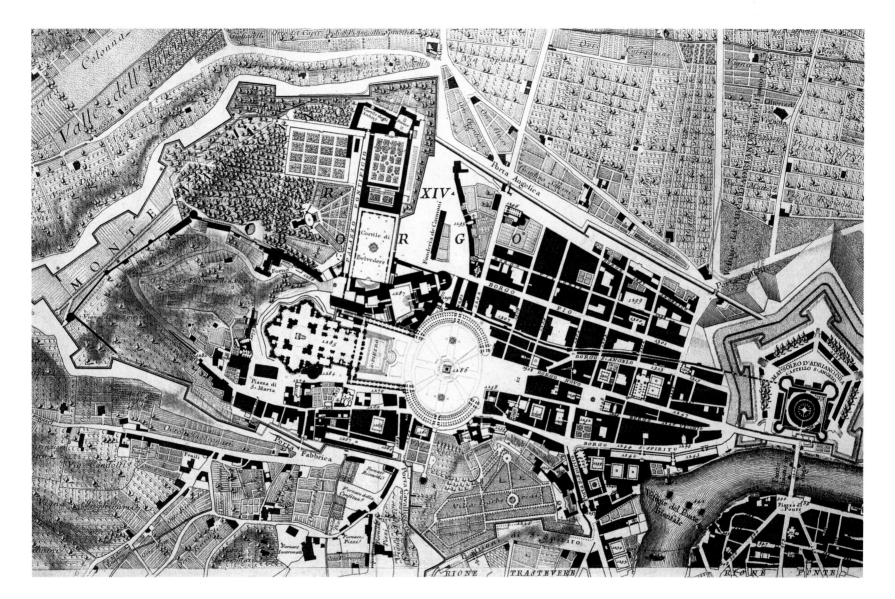

Rome 1748

Above: Part of a set of 12 engraved sheets providing a detailed map of Rome, this sheet shows St. Peter's, designed by Bramante and Michelangelo, and the Castell Sant'Angelo. The plan is dedicated to Benedict XIV, pope 1740–58, who wrote a treatise on canonisation — *De Servorum Dei Beatificatione et Beatorum Canonizatione* — and was also responsible for removing the ban on Copernicanism. The papacy saw much of its power, both political and religious, eroded in the 18th century thanks to Ultramontanism — the trend towards nationalism within Catholic countries.

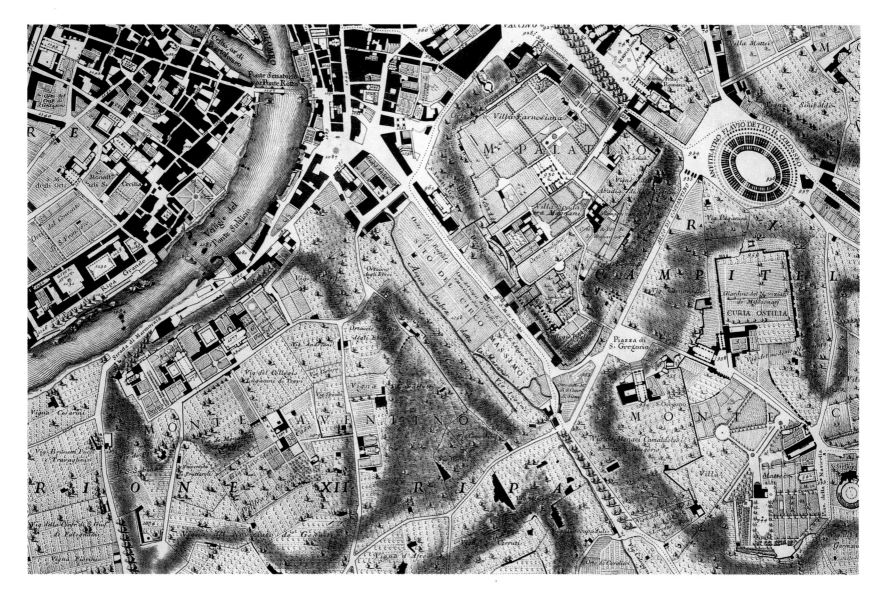

Rome 1748

Above: Another of the set of 12 maps of Rome, engraved by Rocco and Stefano Pozzi, Pietro Campana and Carlo Nolli. This sheet shows some of the great sites of antiquity — the Flavian amphitheatre (the Colosseum), the site of the Circus Maximus and the Palatine Hill.

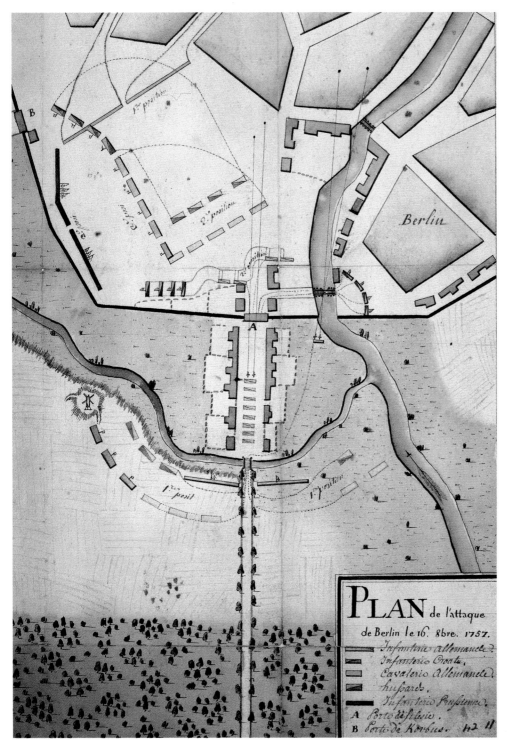

Berlin 1757

Left: The Seven Years' War (1756–63) was a complicated business which saw Frederick the Great of Prussia take on all comers in Europe — the Habsburgs, the French, the Russians, the Swedes and, latterly, the Spanish — while his supposed ally Britain fought it out with France for overseas possessions. There's no doubt that Britain was the winner in the latter conquest: victory at Quebec (1759) secured Canada for the crown and success at Plassey (thanks to the French artillery not keeping its powder dry) led to British ascendancy in India. In Europe, Frederick showed his prowess as one of history's greatest generals in a series of victories, generally with fewer troops, including the shattering victory of Rossbach where 22,000 Prussians defeated the 68,000-strong Austro-French Army. However, Frederick didn't have everything his own way. On 16 October 1757, 3,500 Austrians, under the command of Hungarian general Count André de Hadik, succeeded in getting through to Berlin and occupying the city. Peace came finally at the Treaty of Hubertusberg in February 1763.

France 1756

Right: This and the two maps following on pages 64 and 65 are from a remarkable survey of France — 'levée par ordre du Roy' — comprising 181 sheets, 51.5cm x 65cm in size, and dated 1756–65, roughly the same period as the Seven Years' War. The king who ordered the survey was Louis XV (1710–74), great-grandson of Le Roi Soleil. The first map (right) is of the environs of Paris. Note Versailles, to the south-west of the city, and the boundaries of Paris itself. The city would expand considerably in the following 50 years and see major rebuilding at the hands of Napoleon.

Page 64: This sheet shows Languedoc: stretching from Montpellier in the west to Arles, the Rhône and the Camargue in the east. Note in the middle of the map Aigues Mortes, the port created by Louis IX — Saint Louis — to launch his crusade in 1248.

Page 65: Perigord — with the River Dordogne running across the centre of the map to its confluence with the Vézere. Note the lack of contours on these maps — they had first been used by Samuel Cruquius in 1730, but most mapmakers of the period used shading to represent high ground.

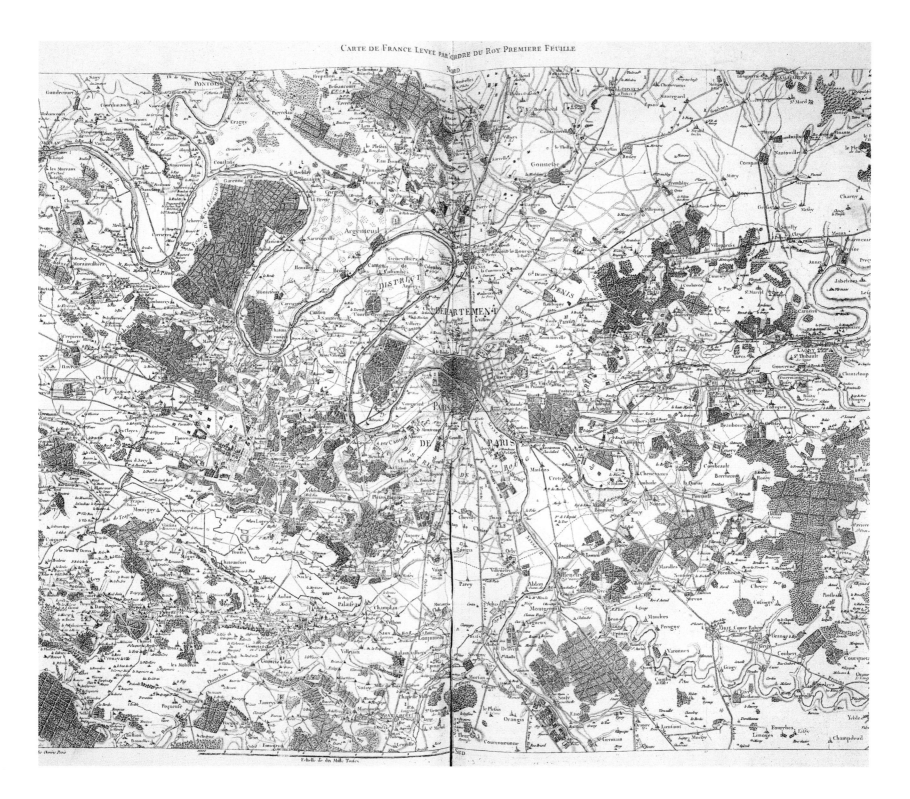

63

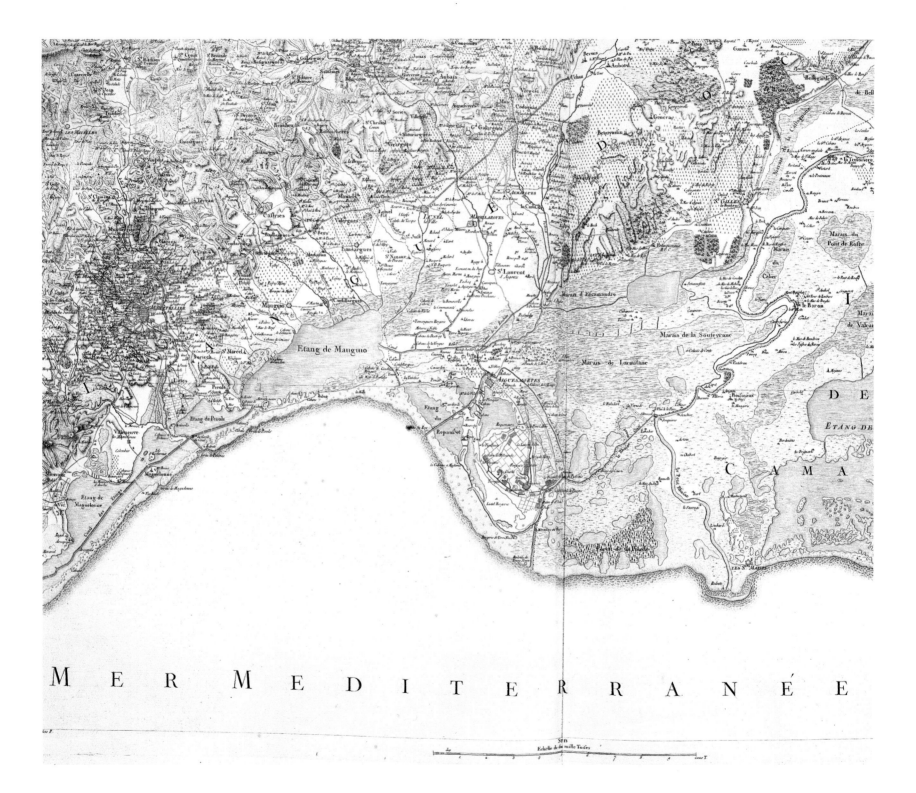

MER MEDITERRANÉE

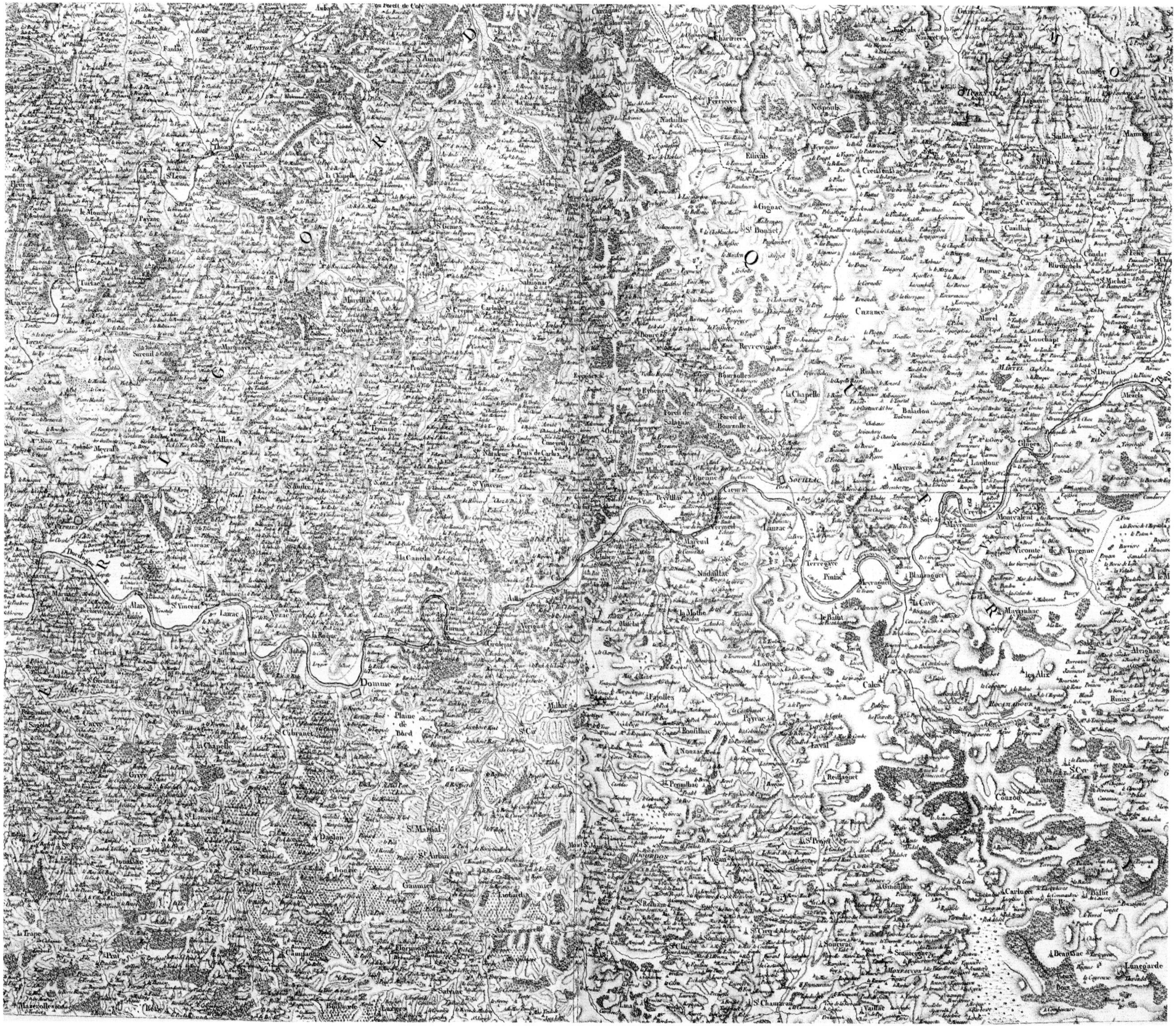

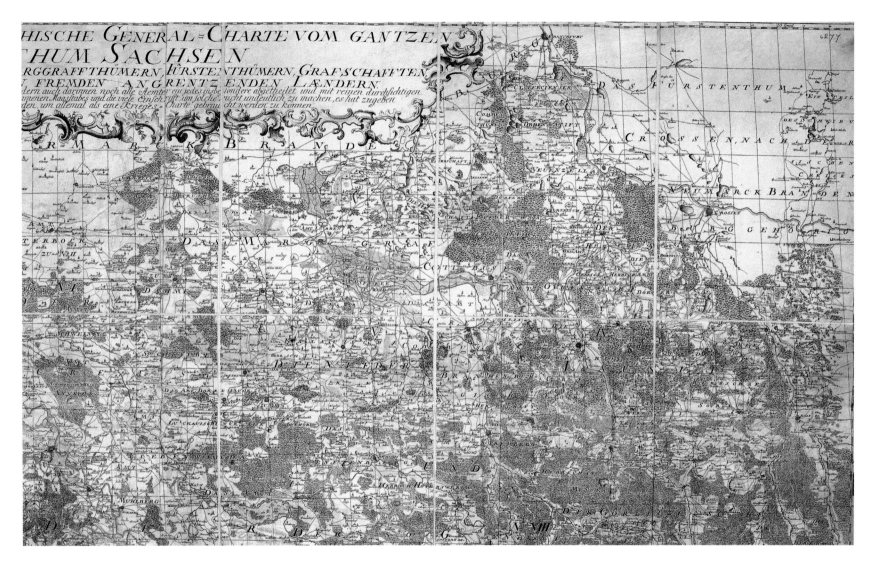

Central Europe 1763

Above and Right: Prepared and drawn by Captain-Lieutenant Petri, Imperial Prussian Engineer, during the Seven Years' War, this engraved map shows Saxony, and parts of Poland and Bohemia (today's Czech Republic), starting just south of Berlin. Interesting locations include Torgau on the Elbe — made famous as the location where Russian and American troops met at the end of World War II — Dresden and Prague. Note that the Frankfurt at top right (above) is on the Oder, and is not the more westerly Frankfurt am Main.

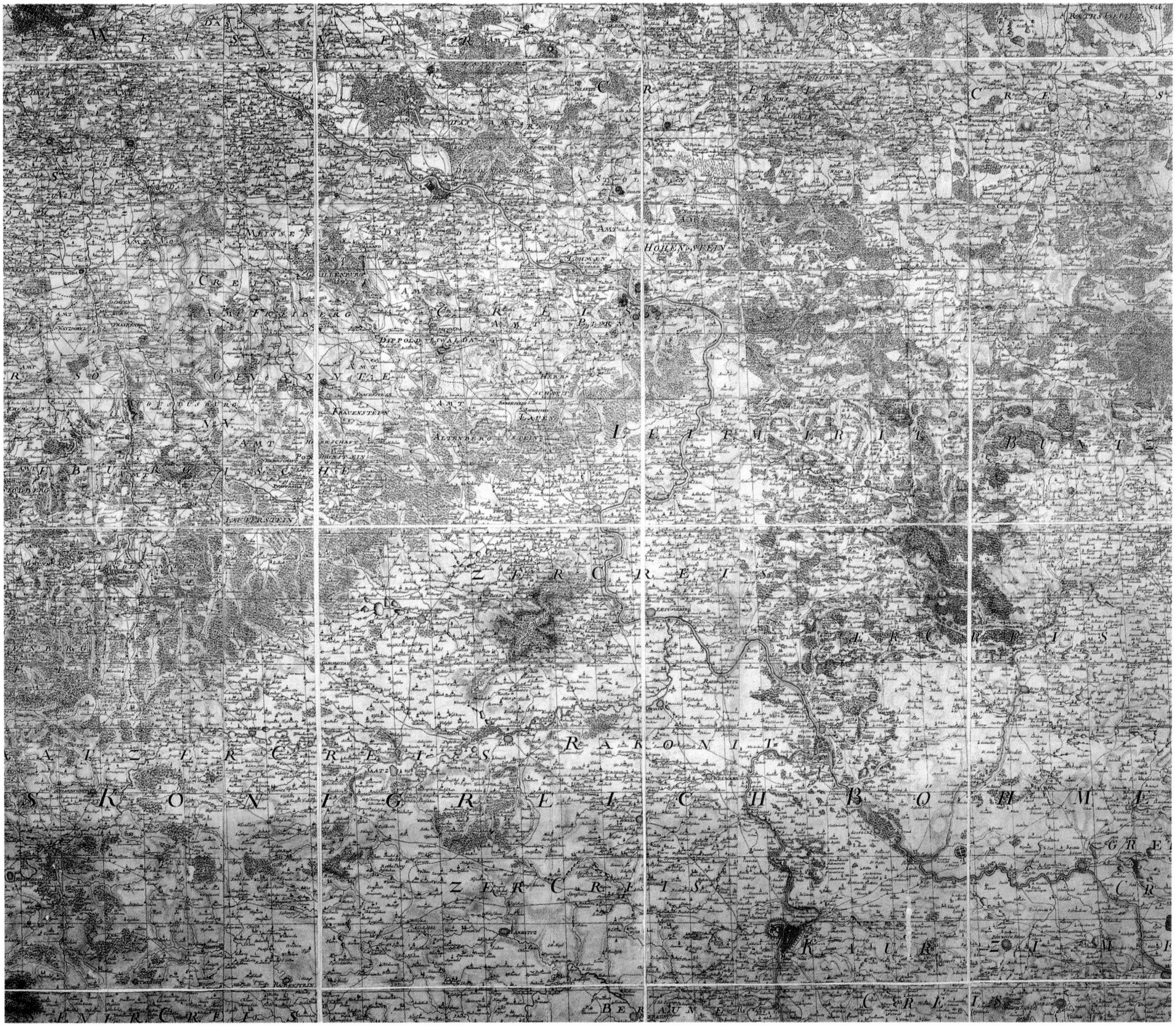
AMT HOHENSTEIN
FRAUENSTEIN
ALTENBERG

PARTITIONS OF POLAND

Partitions of POLAND

1772	1793	1795	
			RUSSIA
			PRUSSIA
			AUSTRIA

Boundary of POLAND in 1772 ....

Limit of the Kingdom of POLAND fixed by the Congress of Vienna (Congress Poland)

G.S., G.S., Nº 2888.

Drawn and printed at the War Office, Nov 1918.

Scale 1:4,000,000 or 1 Inch to 63·13 Miles

Miles 100   50   0   100   200 Miles

Kilometres 100   50   0   100   200   300 Kilometres

The Partition of Poland 1658–1815

The three maps on this spread all show much the same thing: the way that countries were dismembered by the European superpowers of the day — in the case of Poland (left and right) mainly Prussia, Russia and the Habsburgs. Traditionally an enemy of Poland, Russia swallowed the largest portion and was able to take more from the Prussians and Habsburgs as the 18th century progressed. Finally, at the end of the Congress of Vienna in 1815, 26 interested parties shared the spoils after Napoleon's defeat, and more of what had been Poland changed hands.

Left: This map shows graphically the split between Russia (pink), Prussia (green), and the Habsburgs (Yellow) in 1772, 1793 and 1795.

Above: Published by Edward Stanford in London, 27 April 1863, this printed and coloured map of the kingdom of Poland shows by colour and date the successive seizures of its territory by Sweden, Russia, Austria and Prussia.

Above: This map of Europe between The Hague and Constantinople illustrates just what territories changed hands, and to whom they went, at the Vienna Congress Treaty of 1815. Drawn and printed at the War Office in 1918, it shows the frontiers in 1792, territories which changed hands in 1815, new frontiers, and states to which territories were transferred.

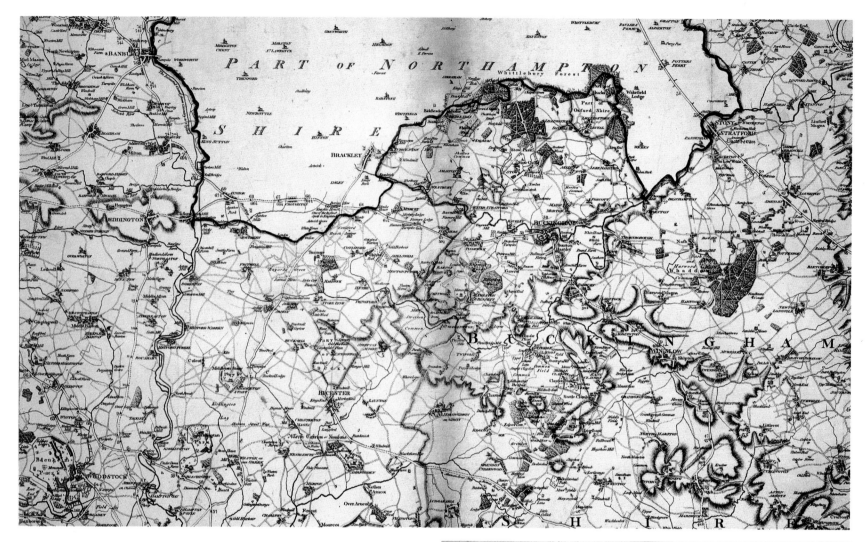

Home Counties, England 1774–76

Above and Right: Taken from 'A Map of the Country Sixty-Five Miles Around London' by John Andre and published by John Stoke these maps make an interesting comparison with the representational approach shown on page 71. The maps have a grid based on degrees of longitude west from St. Paul's Cathedral and make a determined effort to show the main physical features. That of Oxfordshire and Buckinghamshire, from Woodstock to Aylesbury, was published on 10 June 1774; the sliver of Essex (right) came out two years later to the day.

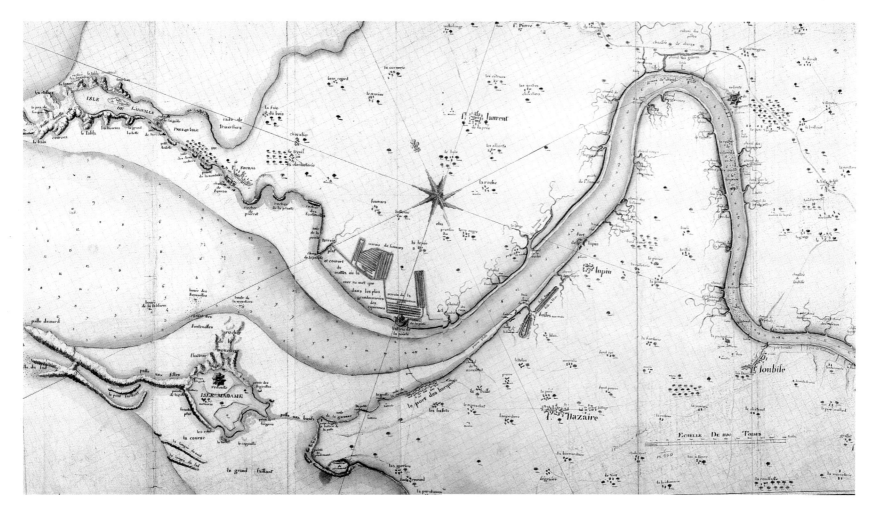

River Charente 1775

Above and Right: This is a plan of the River Charente from Rochefort to the mouth of the river near to Île Madame, with soundings in feet. Drawn by a student from the royal school at Rochefort it was mapped out by M. Digard de Kergüette, 'ancien ingenieur du Roi'.

Switzerland 1782

Above: One of a series of maps of Switzerland, this representational coverage centres on Lake Leman with Geneva at the left tip of the lake and Lausanne at the top centre. The series of maps was dedicated to King Louis XVI, who ruled France between 1774 and his execution on 21 January 1793. Both cities on the lake were major centres of Calvinism, but coexisted with their neighbours after confessional coexistence was recognised by the Treaty of Aarau in 1712. The Swiss would be shaken out of their somnolent century in 1798 when Napoleon invaded.

Kingdom of the Two Sicilies 1783

Left: 'An Accurate Map of the Two Sicilies particularly showing the places destroyed by the late earthquake'. With a key that includes Ancient Roman miles, this map shows an area of Europe that saw much change of leadership over the period as Austrian Habsburgs, Spanish Habsburgs, Bourbons and the House of Savoy all had their moments in charge. Plagued by bandits then as now, Sicily was densely populated but strikingly poor thanks to the rapacious middle class entrepreneurs who took advantage of absentee landlords.

Europe 1783–1807

Right and Details: This six-sheet engraved map (with details on pages 75–77) shows the frontiers of Europe's main powers following the treaties of Versailles 1783, Amiens 1802, Tilsit 1807, and Pressburg 1805. The Treaty of Versailles — signed by Britain, France, Spain and the United States — recognised the independence of the 13 mainland colonies in North America among other issues; Spain recovered Minorca. Britain's influence and prestige in 1783 were perhaps lower than at any other moment in modern history. Following the Treaty of Amiens, which ended the war between Britain and France, Malta, which had recently been retaken by the British from Napoleon, was to be restored — not to France, but to the Knights of St. John. The gains that France made after the Peace of Amiens were unsettling to the minds of other European statesmen and the peace was to last less than two years. The Treaty of Tilsit was signed by Russia, Prussia and France after Russia made an armistice with France mainly because of Tsar Alexander's hatred of Great Britain. The part played by Prussia in the treaty was humiliating in the extreme. Confiscated Prussian provinces were, with Hesse Cassel, to form a new Kingdom of Westphalia and the greater part of the Prussian territories of Poland was to form a duchy of Warsaw. Prussia lost nearly half her territory and population. When Napoleon signed the Peace of Pressburg with the Austrians he inserted provisions for further aggrandisement of the independent German states, mostly at the expense of Austria. The dismantling of the Holy Roman Empire continued between 1805 and 1806.

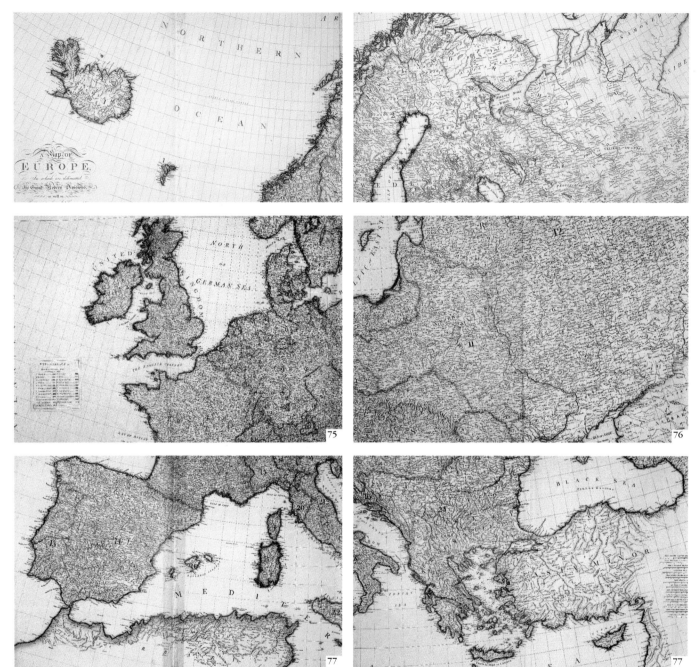

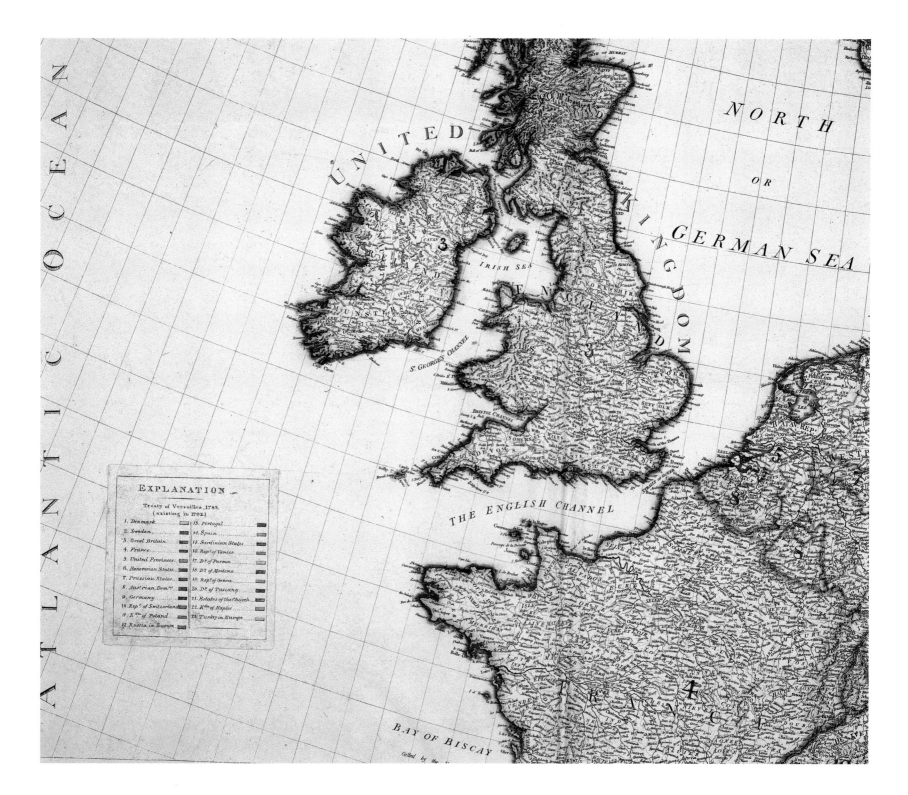

ATLANTIC OCEAN

NORTH

OR

GERMAN SEA

UNITED KINGDOM

IRISH SEA

THE ENGLISH CHANNEL

ST GEORGE'S CHANNEL

BRISTOL CHANNEL

BAY OF BISCAY

FRANCE

EXPLANATION

Treaty of Versailles 1783.
(existing in 1792.)

1. Denmark
2. Sweden
3. Great Britain
4. France
5. United Provinces
6. Hanoverian States
7. Prussian States
8. Austrian Dom.ns
9. Germany
10. Rep.c of Switzerland
11. K.dm of Poland
12. Russia in Europe
13. Portugal
14. Spain
15. Sardinian States
16. Rep.c of Venice
17. D.y of Parma
18. D.y of Modena
19. Rep.c of Genoa
20. D.y of Tuscany
21. Estates of the Church
22. K.dm of Naples
23. Turkey in Europe

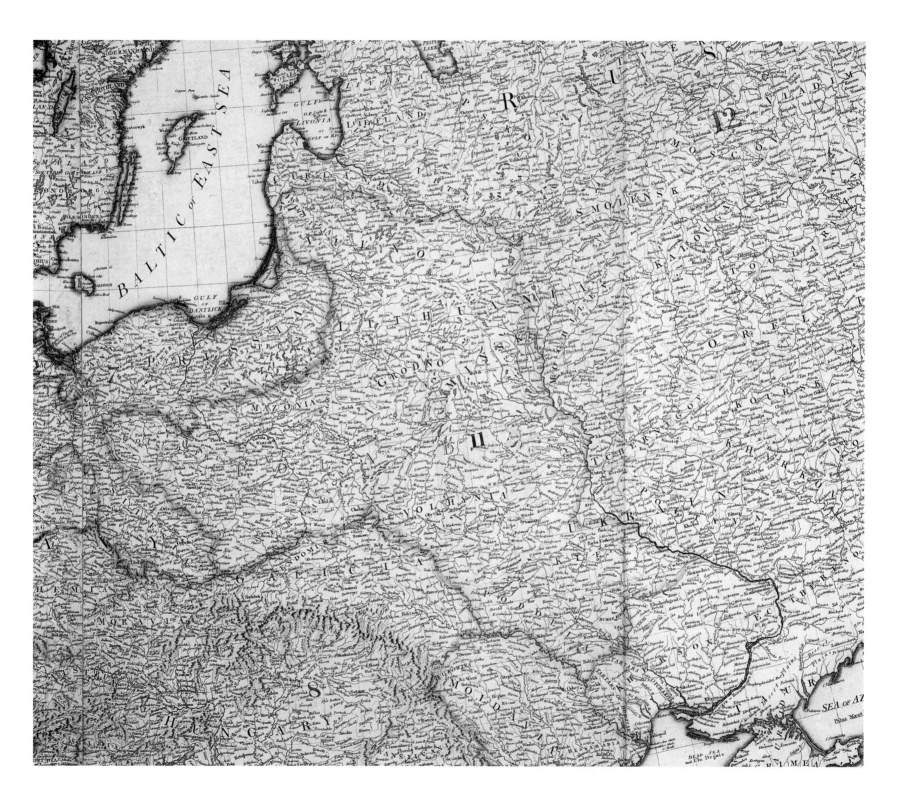

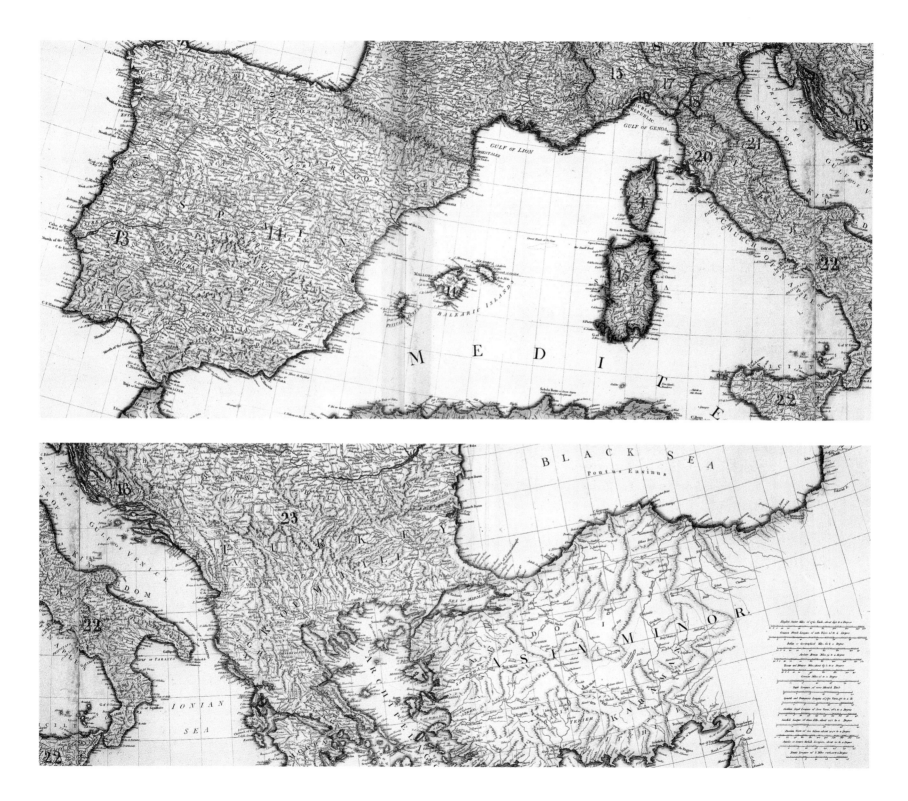

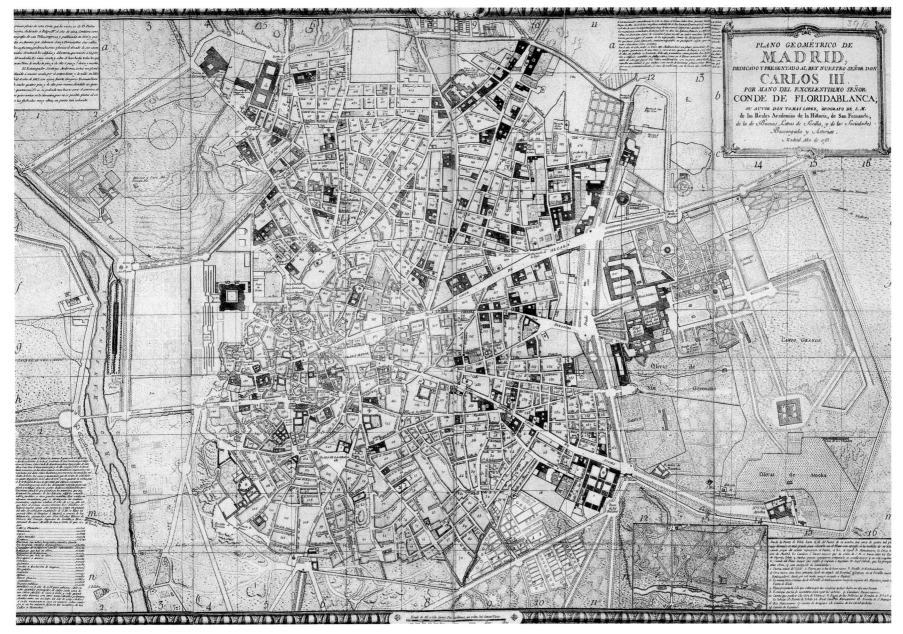

Madrid 1785

Above: This excellent street plan of Madrid was produced during the reign of Charles III (ruled 1759–88) by Don Tomas Lopez. The reference tables are for streets, churches, bridges, convents, hospitals, plazas and colleges. Madrid had had a fairly quiet century, and would continue to do so until 1808, when the city was captured by Napoleon during the Peninsular War.

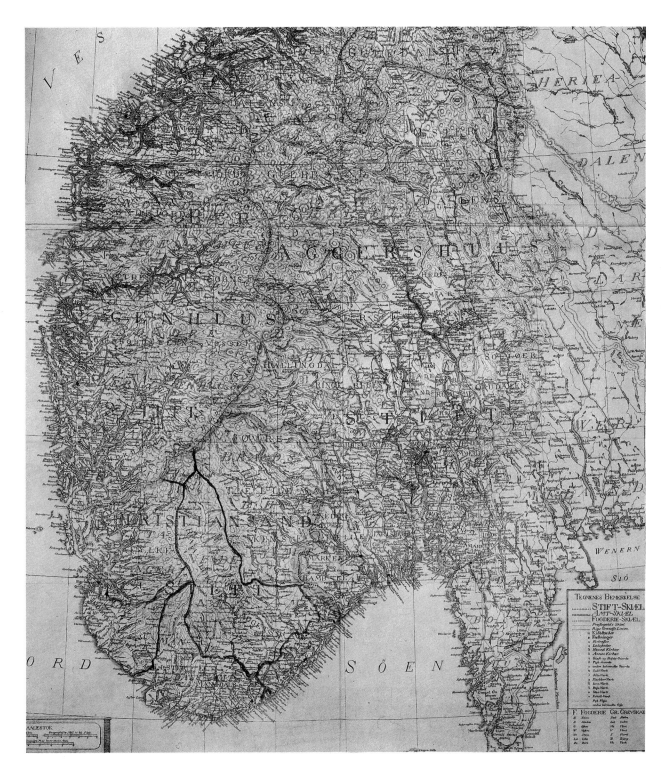

Norway and Sweden 1785

Below and Detail Left: This is a beautiful map of Scandinavia, produced in Norway, with a particularly fine title engraving that includes the Norwegian Arms supported on an obelisk.

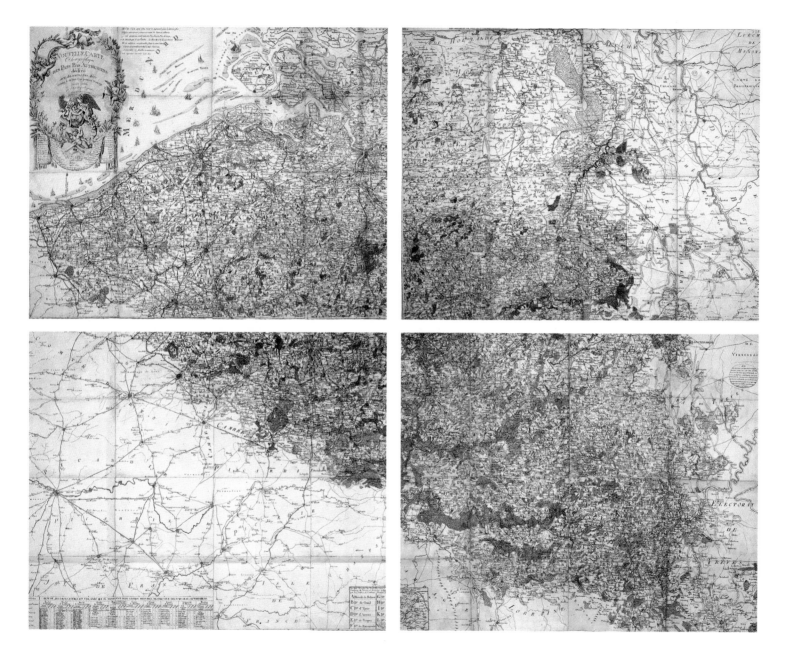

Austrian Netherlands 1789

Above and Detail Right: Following the War of the Spanish Succession, the Austrian Habsburgs took control of the low countries and ruled in an enlightened fashion. This highly detailed engraving shows the Austrian Netherlands, comprising the whole of modern Belgium, with parts of France, Netherlands, Luxembourg and Germany. At right is a detail with the splendid title, and the name of the cartographer (J. B. de Bouge of Brussels, who dedicated his work to the 'amateurs des Arts'). It is interesting to compare the technique of this map to the similar, but 60 years earlier, version on page 49.

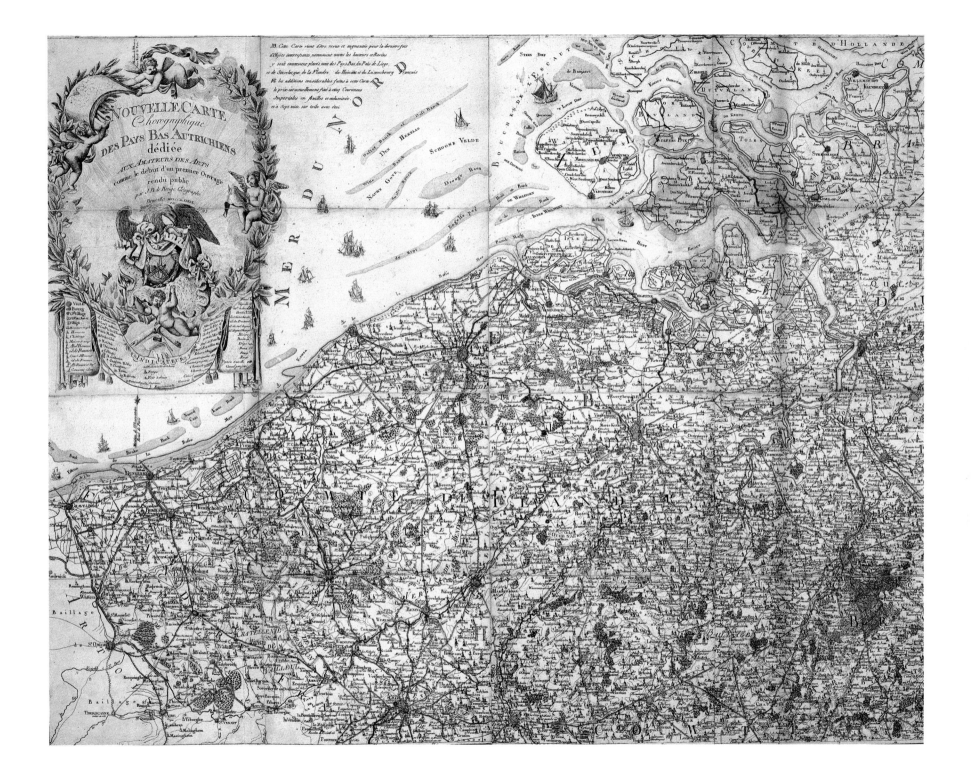

81

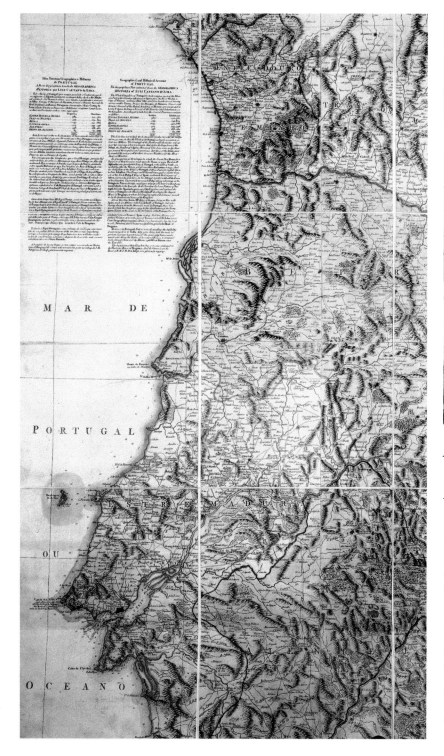

France 1790–91

Right: This and the four maps illustrated on pages 84–87 are taken from the *Atlas National de France*, showing the départements as decreed by the National Assembly in January and February 1790 and on 30 January 1791. Each sheet is engraved, and shows divisions into districts and cantons; rivers, roads and place names. After the French Revolution the old historic provinces of France — Brittany, Normandy, Champagne, Guienne, Burgundy, Provence — were abolished. In their place came 83 départements called after the natural features that belonged to them, without traditions and making no appeal to local sentiment. It was intentionally done, the fine local traditions got in the way of a national unity epitomised by the motto: 'The Republic, One and Indivisible'. The départements illustrated are:

Right: Département de Gard, decreed 3 February 1790, with eight districts and 57 cantons.

Page 84: Département de Pas de Calais, decreed 20 January 1790, with eight districts and 85 cantons.

Page 85: Département de la Moselle, decreed 19 January 1790, with nine districts and 76 cantons.

Page 86: Département des Bouches du Rhône, decreed 9 February 1790, with six districts and 42 cantons.

Page 87: Département du Bas Rhin, decreed 13 January 1790, with four districts and 30 cantons.

Those not illustrated are: Aisne, Basses-Alpes, Hautes-Alpes, Ardennes, Ariège, Aude, Doubs, Hérault, Isère, Jura, Meuse, Nord, Basses-Pyrénées, Hautes-Pyrénées, Pyrénées Orientales, Haut Rhin, Var.

Portugal 1790

Above and Detail Left: Dedicated to John Earl of Bute by Thomas Jefferys, updated and revised in 1790 by Lieutenant-General Rainsford, this map has a dual language history of the country as well as a title that incorporates Britannia and the royal arms of Portugal. Amongst the information provided is the result of the 1732 census (1,742,230 souls excluding the estimated 250,000 ecclesiasticals of both sexes). Mention is made of the long alliance with Britain (since 1660) and the continuation of the line of Braganza (Don Joseph 'now gloriously reigning'). Portugal would prove a significant ally to Britain in the wars against Napoleon at the beginning of the next century.

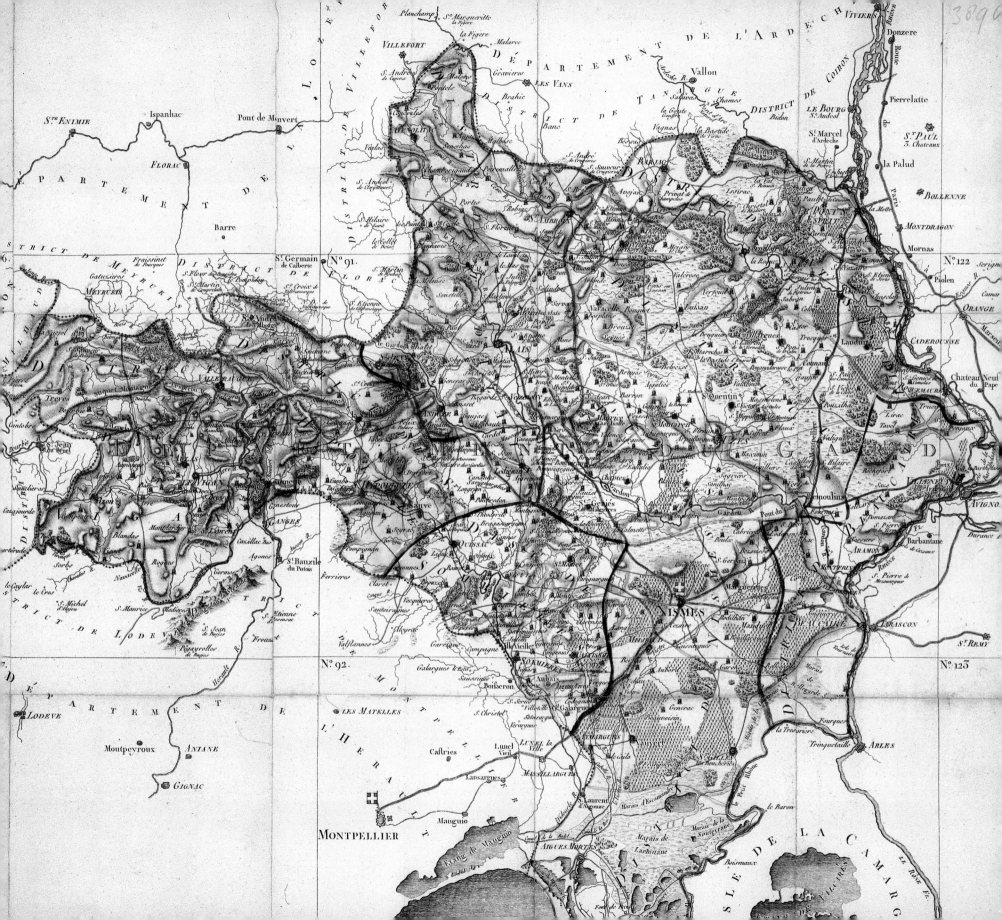

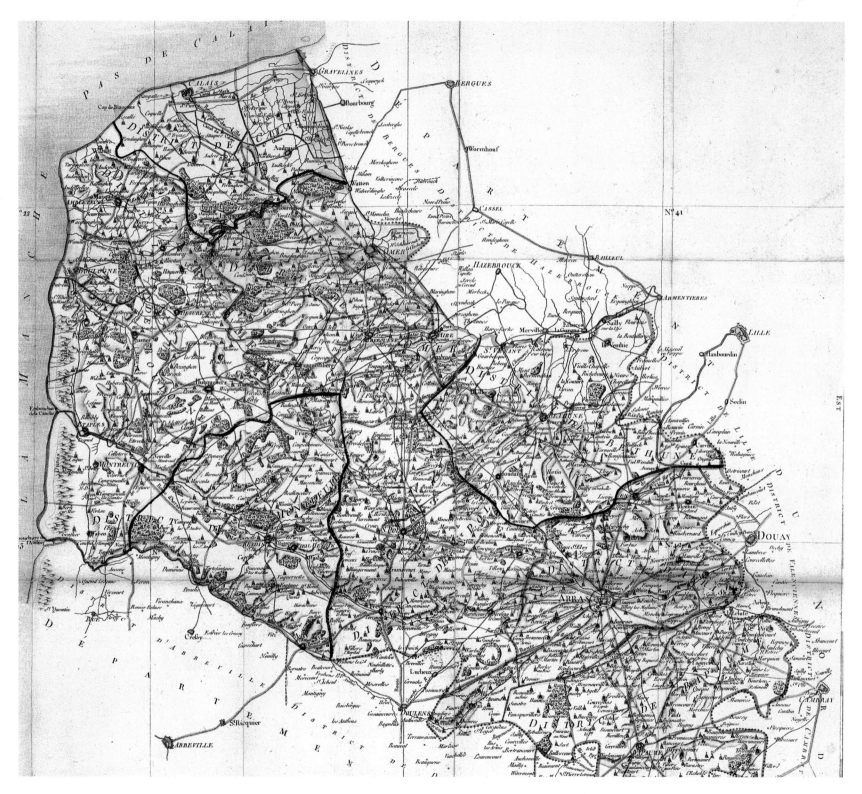

84

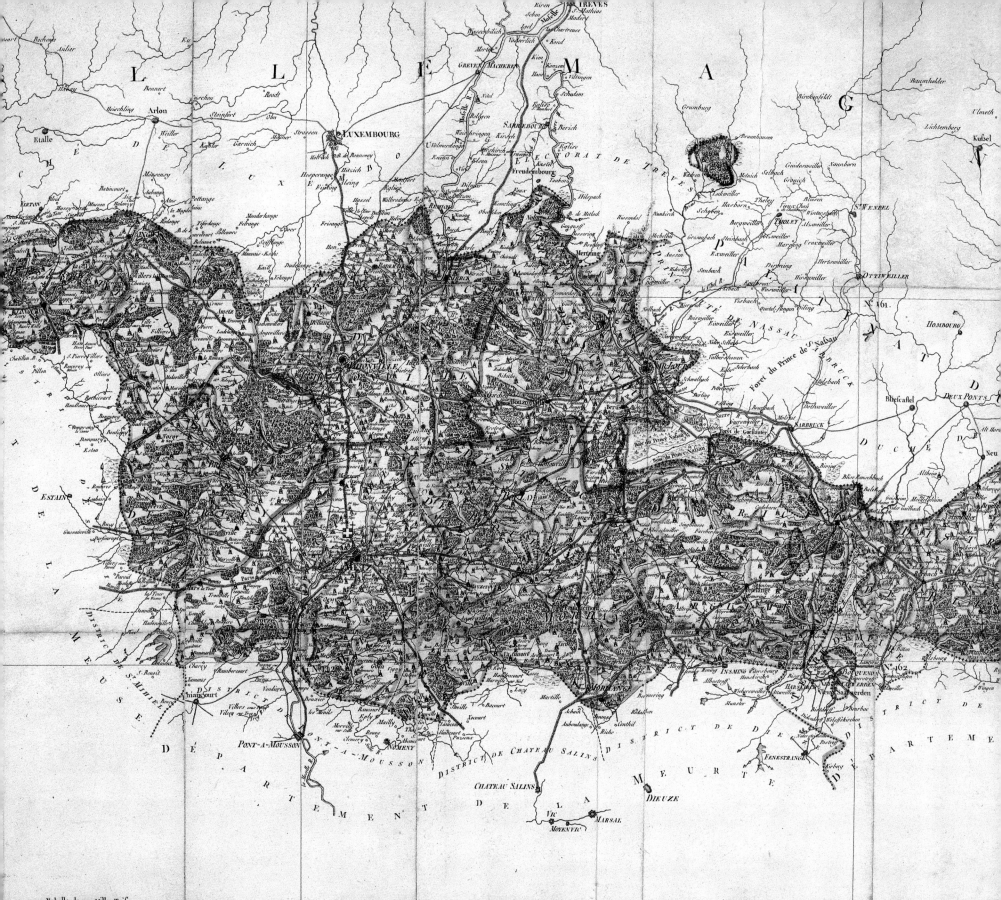

Echelle de 20 Mille Toises.

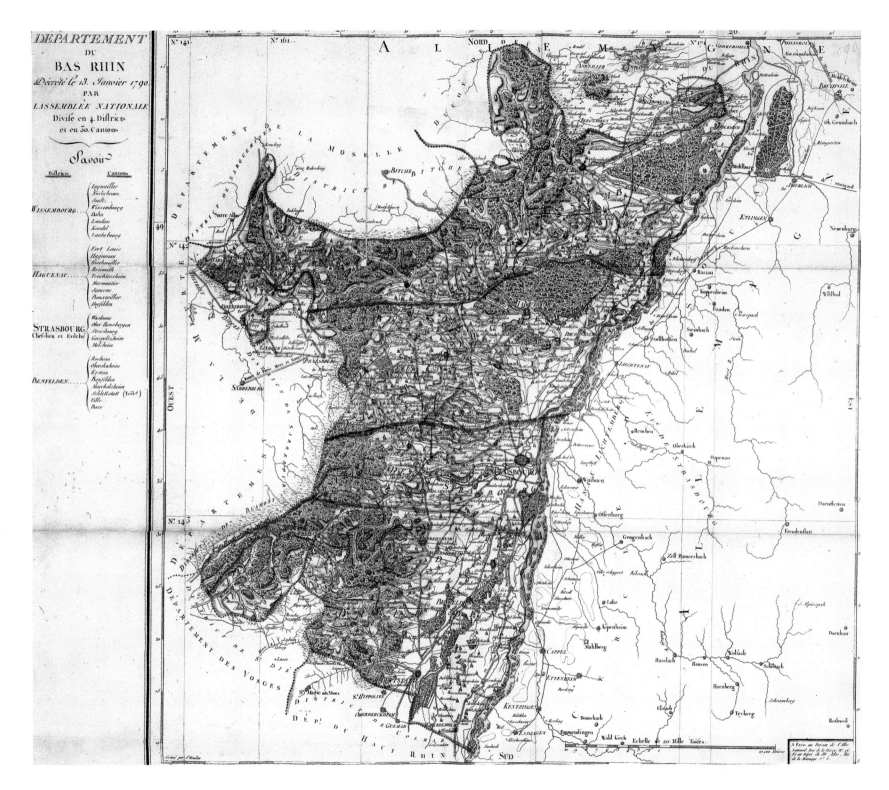

DEPARTEMENT
DU
BAS RHIN
Décrété le 13. Janvier 1790.
PAR
L'ASSEMBLEE NATIONALE
Divisé en 4 Districts
et en 30. Cantons

Savoir

District.	Cantons
WISSEMBOURG...	Ingweiller Niederbronn Soultz Wissembourg Dahn Landau Kandel Lauterbourg
HAGUENAU...	Fort Louis Haguenau Buschweiller Brumath Truchtersheim Marmoutier Sawerne Bouxwiller Bosselden
STRASBOURG. Chef-lieu et Evêché	Wasselonne Ober Hausbergen Strasbourg Geispoltzheim Molsheim
BENFELDEN...	Rosheim Oberehnheim Erstein Benfelden Marckolsheim Schlettstatt (Trib.d) Ville Barr

Echelle de 20 Mille Toises.

87

Italy 1803

Above: An unusual map of Italy and the Dalmatian coast centring on the Gulf of Venice, showing the main roads, rivers and towns but little other topographical detail. By 1803, Napoleon had conquered the peninsula and was instigating his reforms that would eventually see the country divided in three: those parts that were taken by France (including, after 1810, the Papal states); those that became the kingdom of Italy in 1805 (with Napoleon as king); and the kingdom of Naples ruled initially by Napoleon's elder brother Joseph (reigned 1805–08).

Luxembourg City 1801

Right: Coloured plan of Luxembourg City, with a reference table to fortifications, gates, seminaries etc. The style of the plan is similar to dated plans of the late 18th century. At this stage Luxembourg was part of France, having been annexed in 1795, but after the Congress of Vienna it would once again become a Grand Duchy in its own right, albeit with William I, King of the Netherlands its titular head and a Prussian army garrisoned there. It would take until 1867 to throw off the yoke and become an independent nation.

Europe 1800

The next 12 pages (as identified on the overall map to the right) are taken up with Captain Chauchard's highly detailed 'General Map of the Empire of Germany, Holland, the Netherlands, Switzerland, the Grisons, Italy, Sicily, Corsica and Sardinia'. Published and dedicated to King George III (ruled 1760–1820) by John Stockdale, it is a substantial work of reference that is dated 4 June 1800.

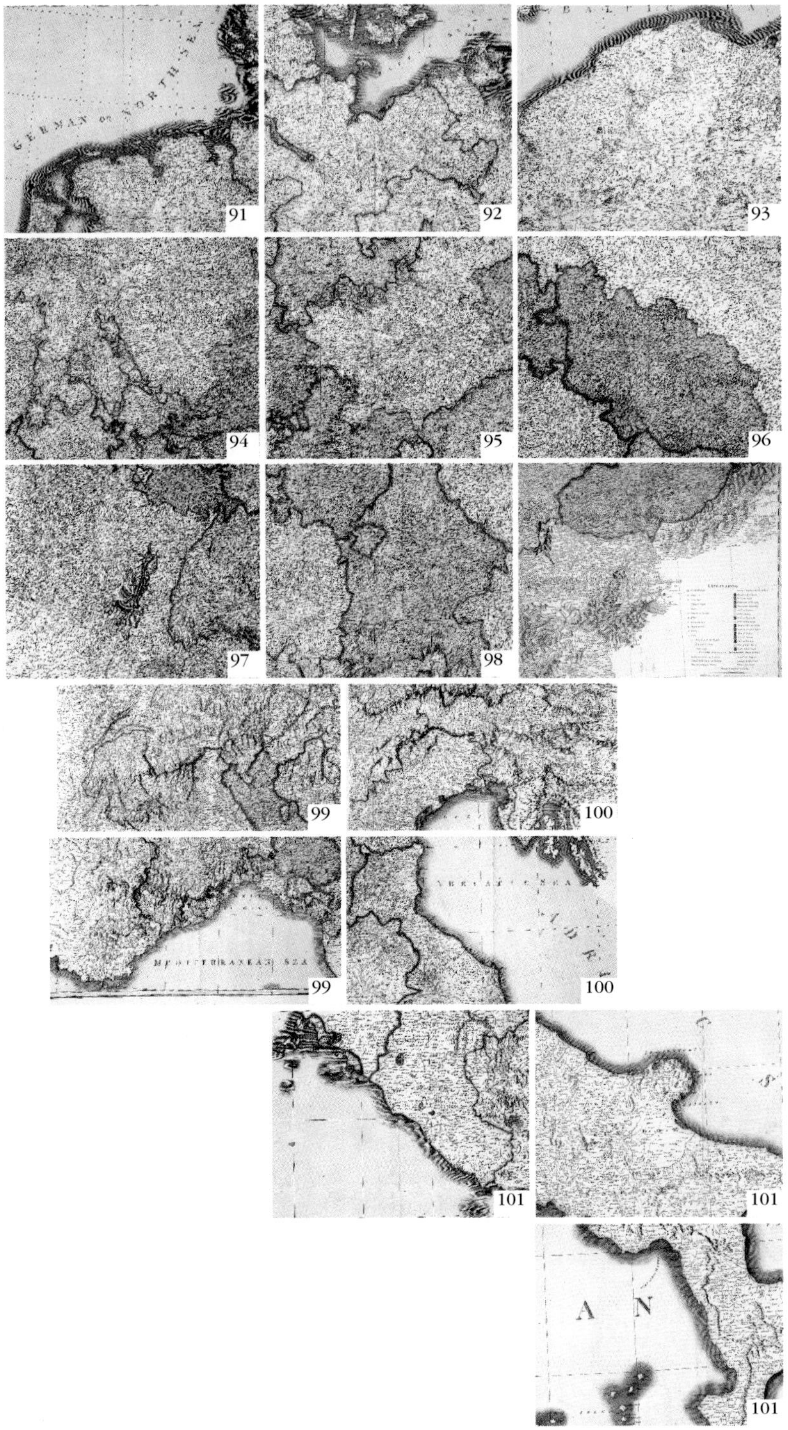

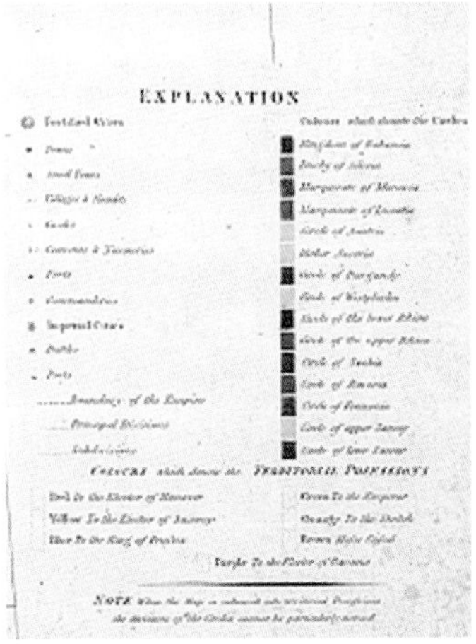

GERMAN OR NORTH SEA

ZUYDER
ZEE

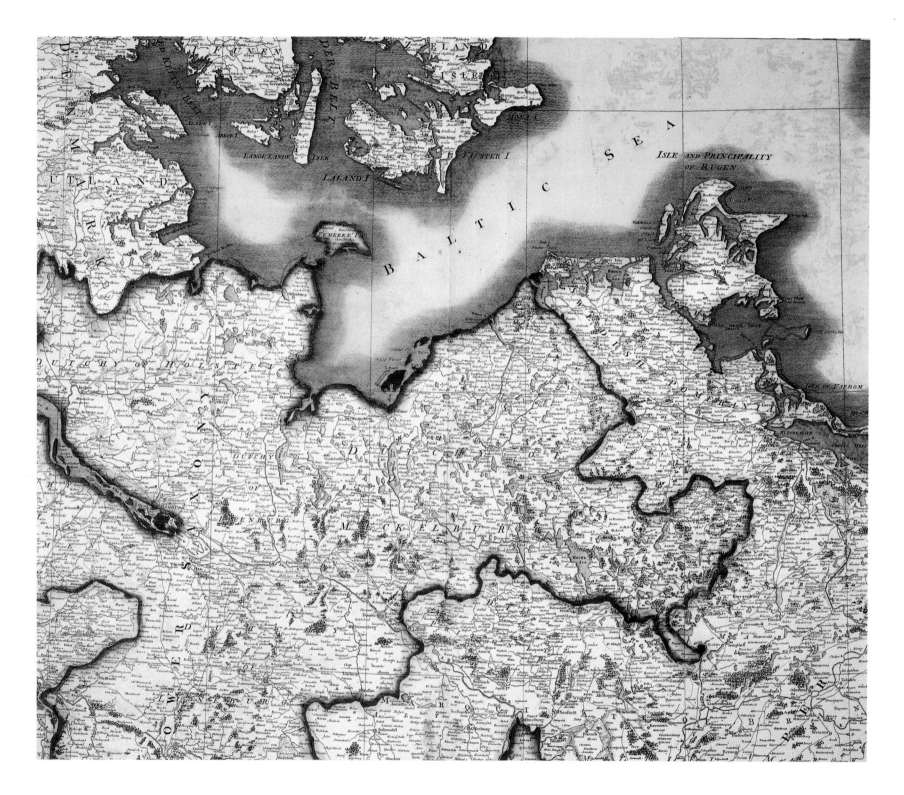

92

BALTIC SEA

GULF OF DANTZIC

BORNHOLM I.

POMERANIA

PART OF PRUSSIA

DANTZIC

KINGDOM OF

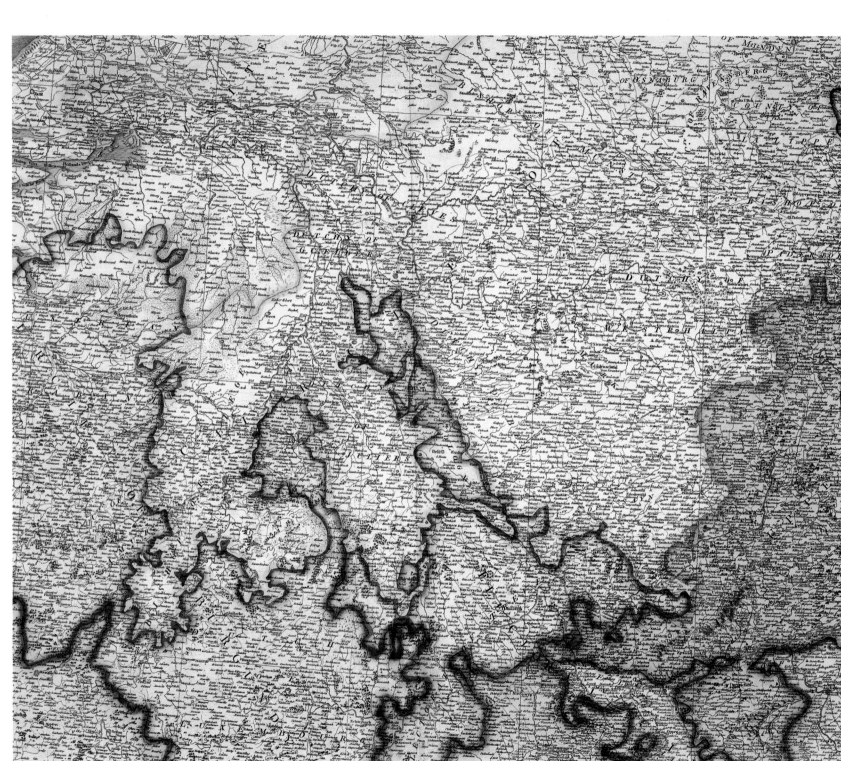

94

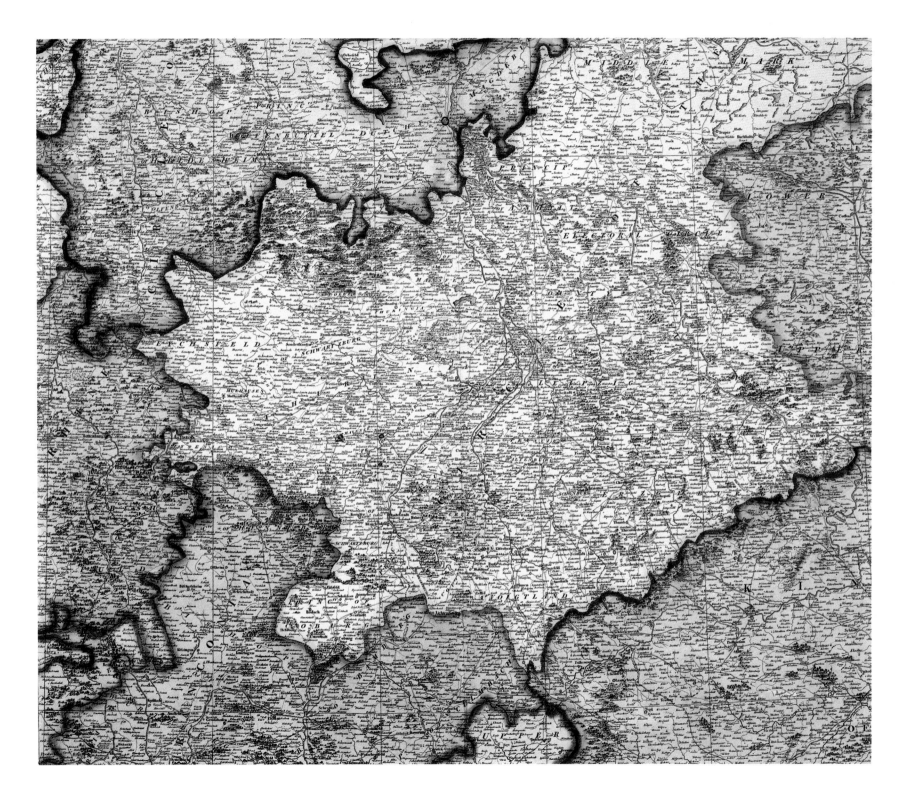

95

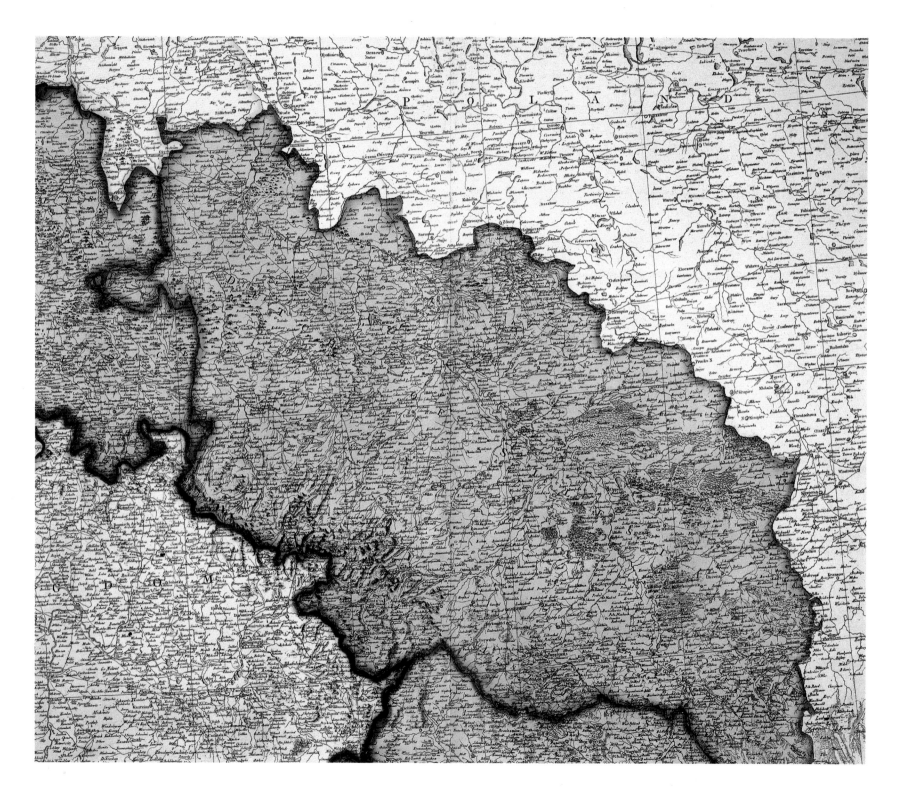

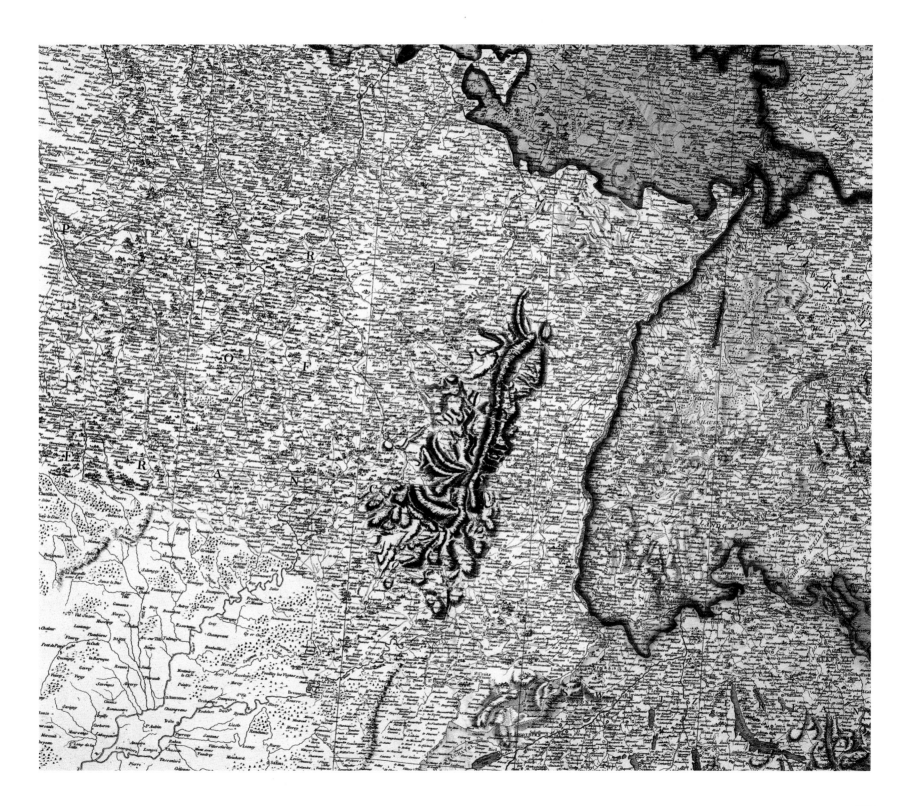

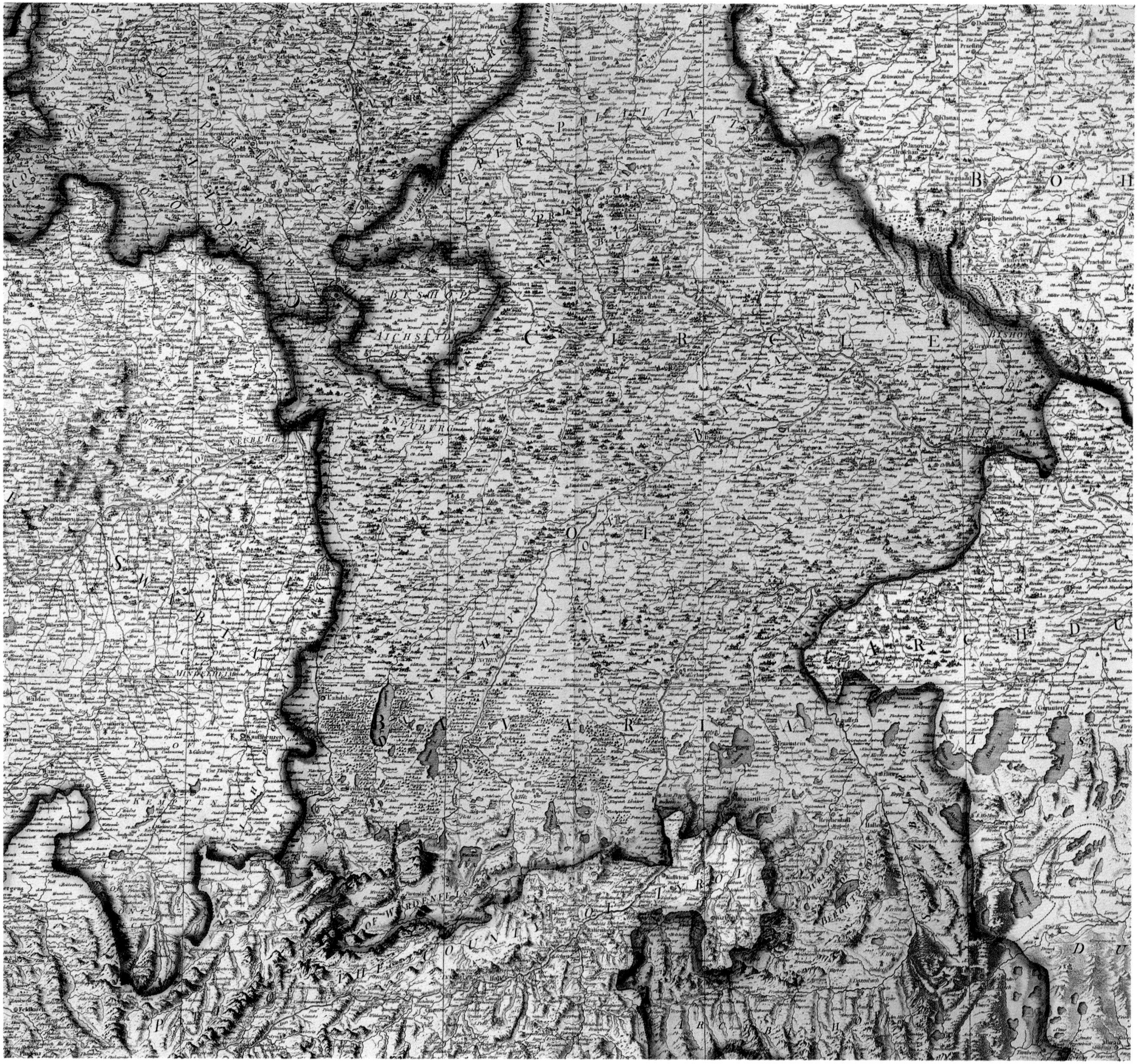

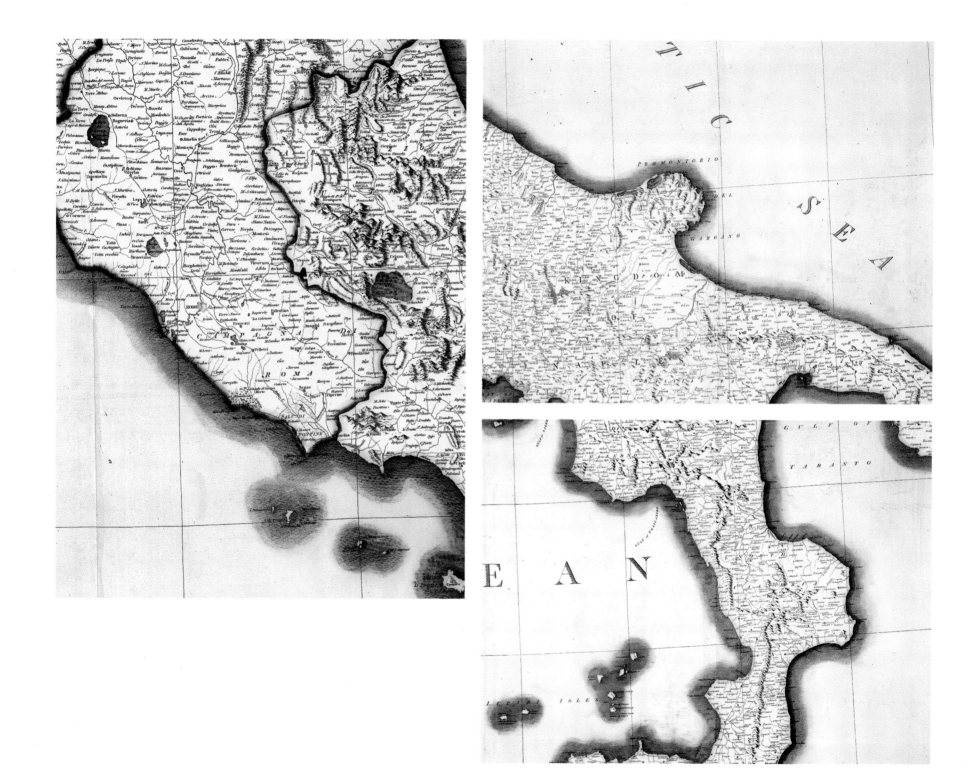

Europe 1805

Above and Above Right: This map, by John Fairburn of London, is titled 'Theatre de la Guerre' and purports to show 'the Scene of Action on the German, French, Dutch, Swiss and Italian Frontiers Exhibiting at One View the whole Theatre of War from the Mediterranean Sea to Dutch Brabant and Gelders, including the Rivers Rhine, Po, Danube, Neckar, Mayne, Lahn, Moselle, Meuse, &c. &c.'. It was published on 21 October 1805 — the very same day of Nelson's famous victory over the Franco–Spanish fleet at the Battle of Trafalgar — and shows the rapidly expanding control of the self-proclaimed Emperor Napoleon that caused Britain to declare war in 1803. The map was found among the papers of William Pitt the Younger (1759–1806) who was appointed prime minister of Britain at the age of only 24 in 1783. After a long political career he died in office after hearing of Napoleon's victory over Britain's allies Russia and Austria at Austerlitz.

North Italy 1798

Right and Far Right: This map is part of a set showing the disposition of forces in North Italy and Switzerland during the campaigns there. The key at right explains the forces involved.

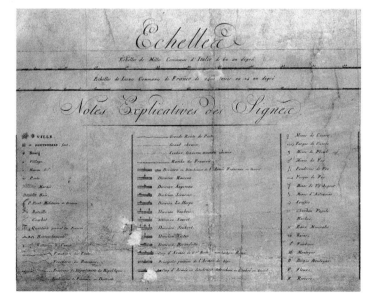

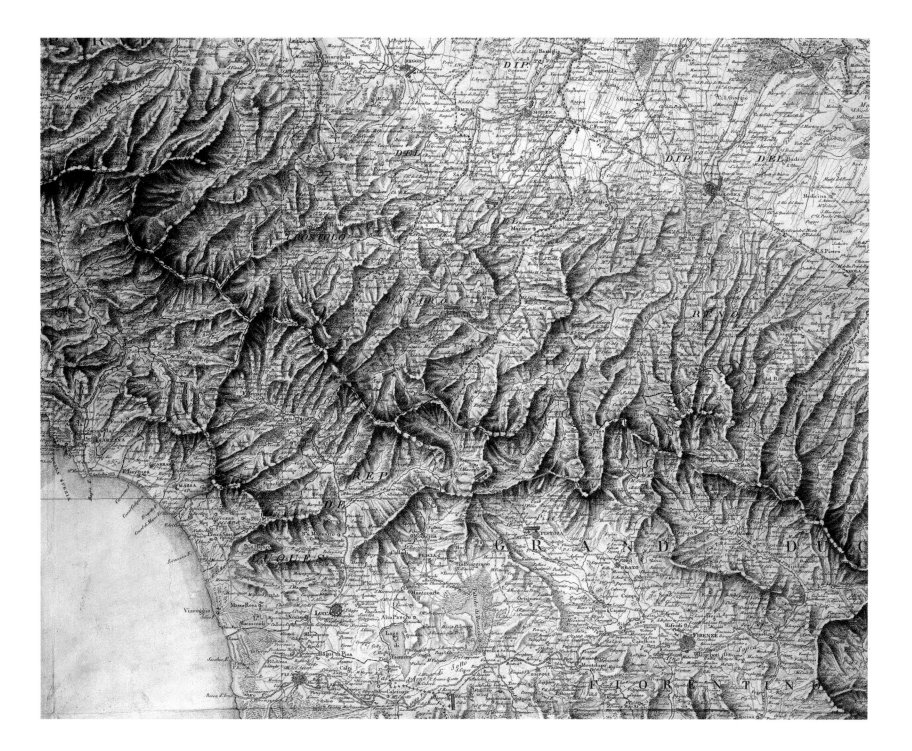

103

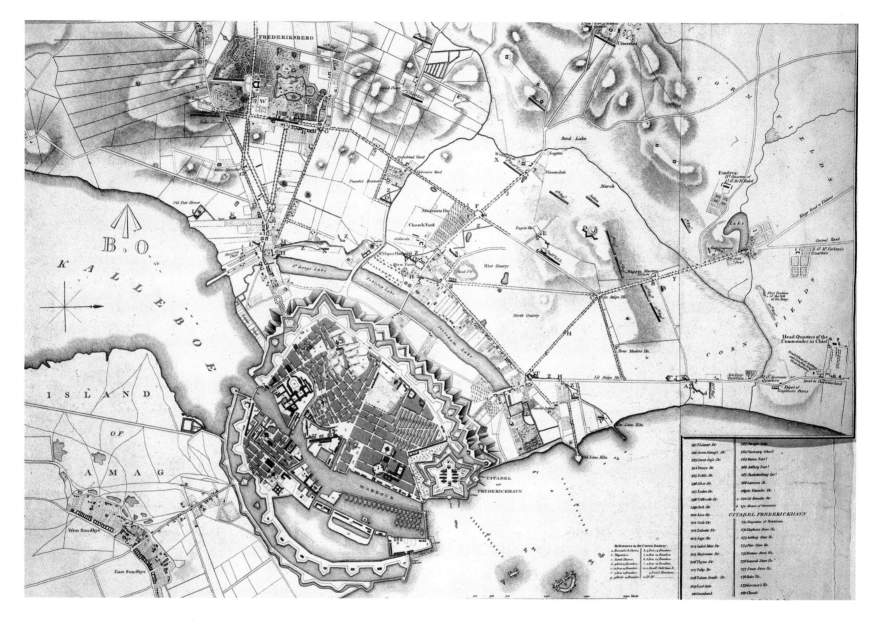

Copenhagen 1807

Above: In 1807, acting on a threat by Denmark to close her ports against British merchandise and join Napoleon in the war against her, the British fleet bombarded Copenhagen and seized the Danish fleet. This engraved plan of the city of Copenhagen shows adjacent ground and the positions of the several batteries erected by the British during the siege in September 1807, which was commanded by Lieutenant General Lord Cathcart. There is a reference table to buildings, streets, batteries, defences, trenches and troops.

Pyrenees–Adriatic 1811

Right: Dedicated to Napoleon, this map details 'the French Empire, Italian Kingdom and states under the protectorate of the Emperor Napoleon'. It was amended in colour to show 'old France' (the 1792 boundaries). Until the withdrawal of Napoleon's troops in 1814 the entire peninsula was under French domination. To the west of Italy were 'département' states while the east was a kingdom ruled by the Bonaparte family. To the south of Italy was the Kingdom of Naples, another vassal state controlled by the Emperor.

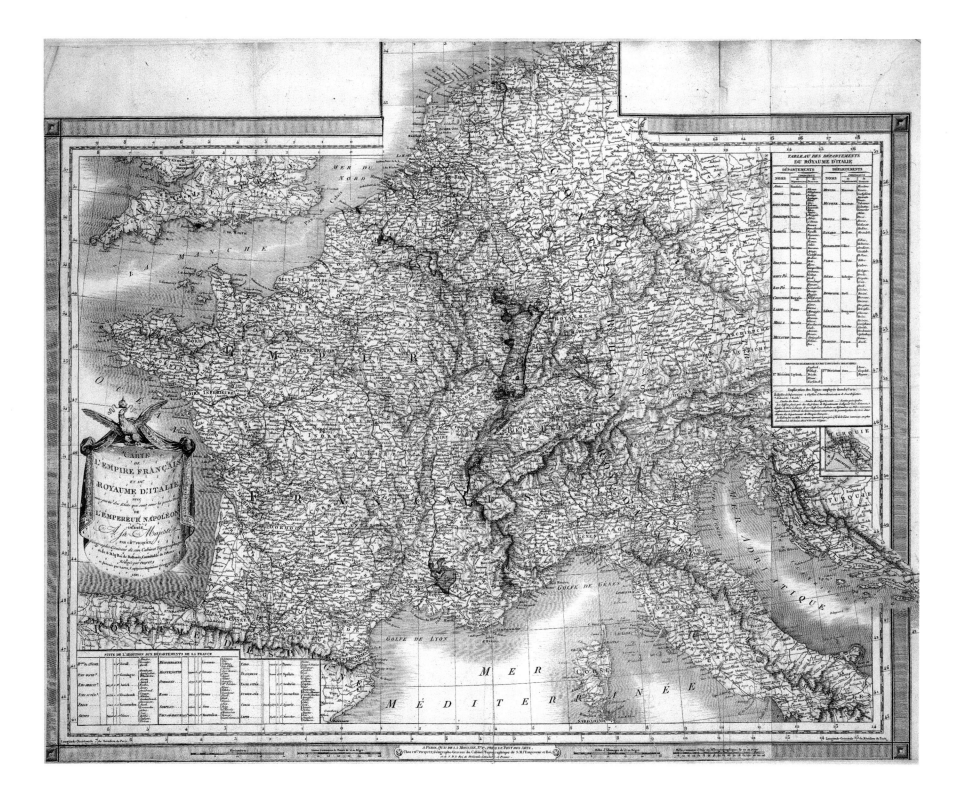

105

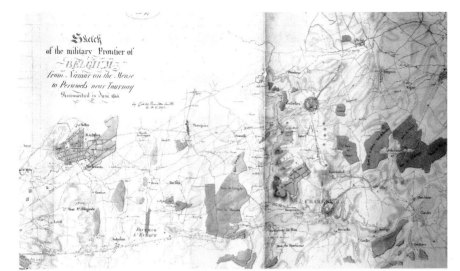

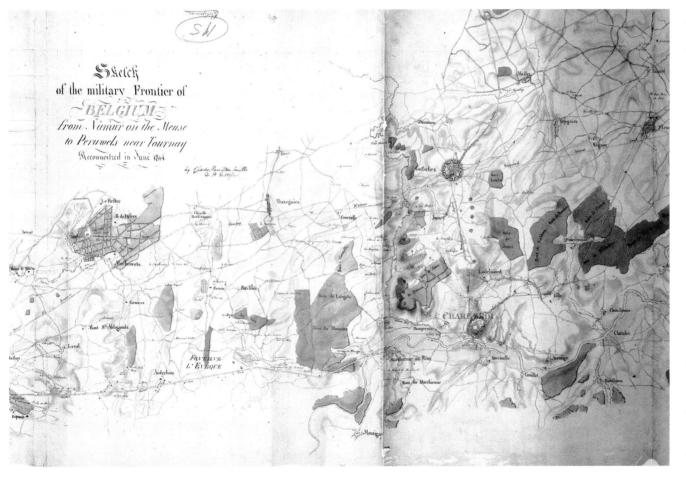

Belgium 1814

Above and Details Left: These four maps (shown in sequence across the top of the two pages) show a 'Sketch of the military Frontier of Belgium from Namur on the Meuse to Peruwels near Tournay, Reconnoitred in June 1814 by Charles Hamilton Smith, Deputy Assistant Quartermaster General'. Smith was born in Flanders of a protestant family and served in the 8th Light Dragoons as a volunteer from 1794. Commissioned lieutenant in 1797, he went on to serve in the West Indies and as deputy quartermaster general on the ill-fated Walcheren expedition of 1809. He served in the Netherlands in 1813–14 and in 1816 undertook an intelligence mission in North America. At the time that Smith was drawing his map, Napoleon was languishing on Elba following the allied invasion of France and his subsequent abdication. However, he would return less than a year later to lead an army that would take on the English, Dutch and Prussian forces just a few miles to north of the area shown here. A little over a century later this area would witness some of the most bitter fighting of World War I.

108

Spain 1770

Far Left: Dated 1770, this map entered the collection of the War Office via Lt-Col Elphinstone in July 1814, by which time the Napoleonic wars in the Iberian Peninsula were over. By Tomas Lopez, it is dedicated to 'Serenissimo Señor Don Carlos Antonio, Principe de Asturias', and shows well the provinces of Spain, Portugal and the Balearic Islands off the Mediterranean coast.

Peninsular War 1808–14

Left and Below: A constant thorn in Napoleon's side — he called it the 'Spanish ulcer'—the Peninsular War would contribute substantially to Napoleon's downfall. He had decided to overthrow the Bourbons in Spain, put his brother Joseph on the throne and invade Portugal; England, Portugal's ally since 1660, came to her aid on 1 August 1808, when Sir Arthur Wellesley — the duke of Wellington — landed in Mondego Bay north of Lisbon. The long years of fighting in Spain and Portugal, so graphically represented by the paintings of Goya, would see many ups and downs: in 1808, while Wellington was in England, the British forces under Sir John Moore, harried by Napoleon, retreated to be evacuated from La Coruña, the peninsula of which is shown illustrated below. Wellington's return saw Portugal defended from attack in 1808–10, a first offensive in Spain 1811–12 and the victorious advance into France in 1813–14. The plan to the left, of the storming of Ciudad Rodrigo in January 1812, shows events that typified much of the fighting in the peninsula.

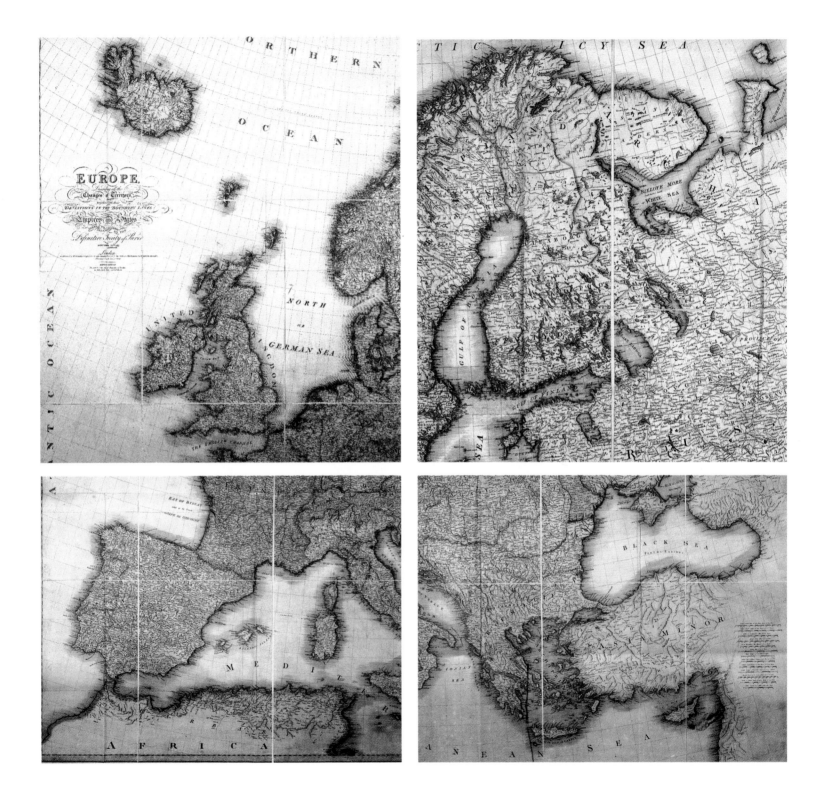

110

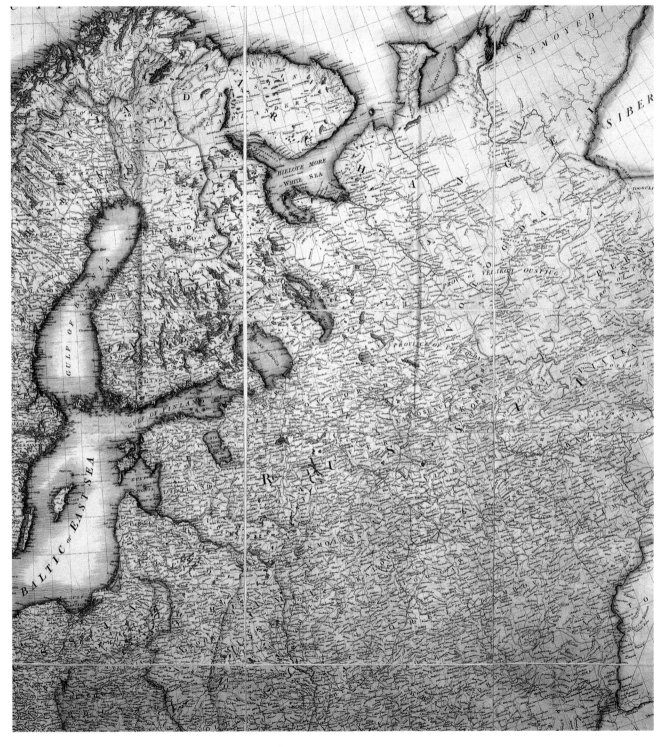

Europe 1815/16

Far Left: This sequence of four engraved, coloured maps (details on pages 111–12) published by William Faden in London, January 1816, describes all the changes of territories in Europe together with the deviations in boundary lines of its several empires and states agreed to by the Treaty of Paris, 20 November 1815, after the fall of Napoleon. This was the second time that such a treaty had been signed; the first, which followed Napoleon's original exile, had been fairly lenient towards the French but this second treaty bound France to submit to the frontiers of 1790. In addition she surrendered Saarbrücken to Prussia, Landau to Bavaria and Savoy to Sardinia and paid reparations of 700 million francs as well as submitting to occupation for a period of five years. Bonaparte suffered perpetual exclusion and his dynasty was prohibited from ascending the throne of France. After Waterloo Napoleon had given himself up to British forces and was exiled to the island of St. Helena in the Atlantic. He died there in 1821.

Russia 1815/16

Left: At the start of the 19th century Russia was the largest country in the world in terms of geography and the end of the Napoleonic wars saw Emperor Alexander I extend his boundaries into Finland and gain a foothold in the Caucasus. The Russian economy also benefitted from the devastation wrought by the invading French. The rebuilding work sparked an entrepreneurial response from the lower classes of the population and marked an economic shift in Russia that quickened the country's industrialisation.

United Kingdom 1815/16

Right: The Act of Union of 1800 dissolved the Irish parliament and forged the union of Great Britain and Ireland — the United Kingdom. From the Napoleonic Wars it gained Cape Colony (South Africa), which had been previously Dutch. This was an important station on the Indian trade route that was starting to bring great wealth to the nation. And with the Industrial Revolution gaining momentum by the day, the scene was set for the country to take centre stage in world politics and carve out an empire that would rule over a quarter of the world's population a century later.

France 1815/16

Far Right: With Bonaparte exiled, a chastened France surrendered her conquered territories and returned Louis XVIII to the throne. However, her revolutionary spirit was not yet vanquished and time and again over the subsequent decades Paris would erupt, threatening to overwhelm the established order of monarchy, Church and property.

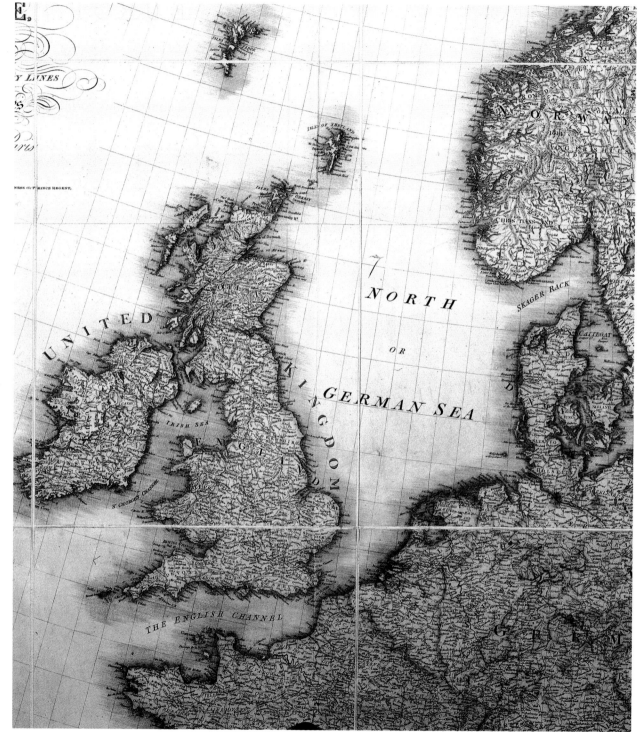

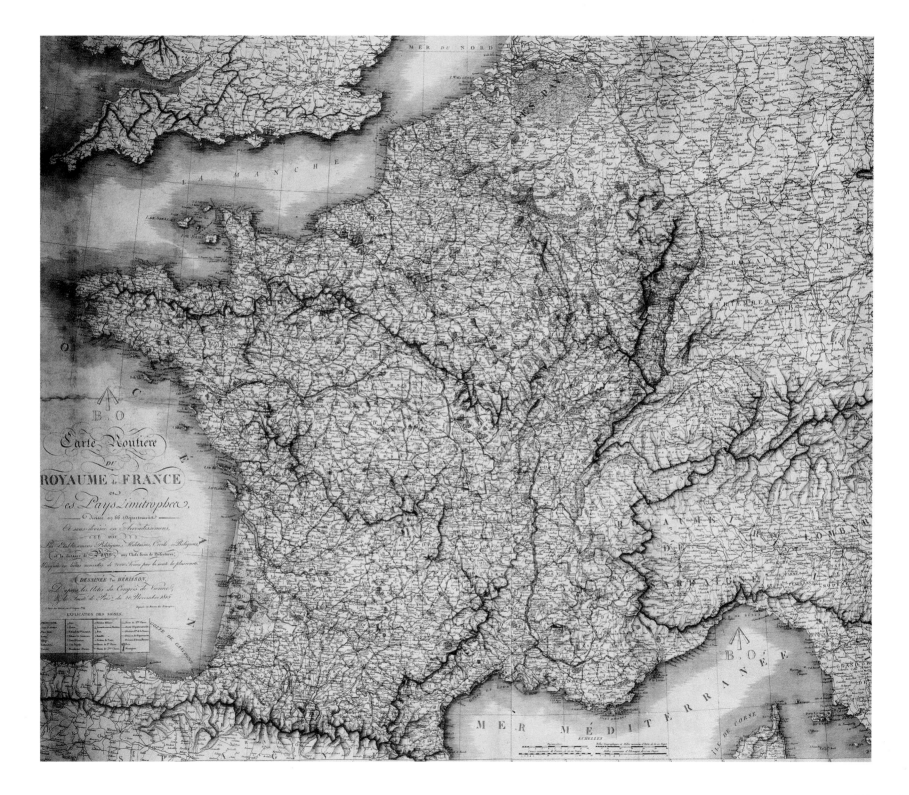

113

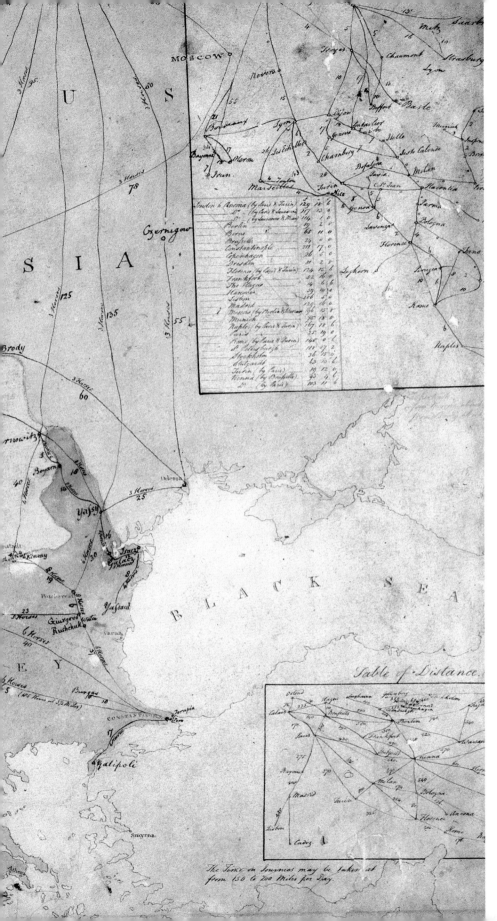

European Courier Routes c.1820

Left: This charming map shows the routes taken by horse couriers throughout Europe. It was not until the 1830s that the steam locomotive started to revolutionise transport in Europe and, while steam engines were beginning to be used in ships, horses were still the only way to travel across land during the 1820s. Information on the map gives distances in English miles and a conversion table of English miles into European units of length. It also shows how much it cost to make the crossings from Dover and Harwich to various ports on the continent.

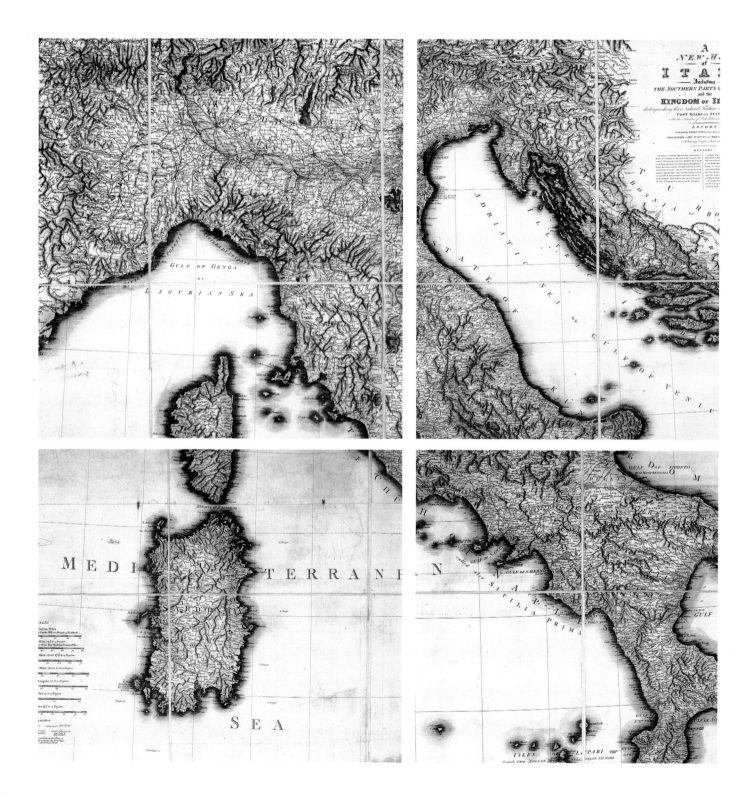

Italy and Neighbours 1825

Far Left and Detail Left: Described as 'A new map of Italy, Including the Southern Parts of Germany, and the Kingdom of Illyria . . .', this map was published by James Wyld, London, on 1 January 1825. It is an engraved, coloured map, showing rivers, mountains, towns, post roads and stations and boundaries. Illyria, to the east of the Adriatic, included the western coastal region of the Balkan peninsula and was a part of the Austrian Empire at the time the map was published.

For a period of almost 50 years after French rule Italy did not exist as a unified country. Until the Kingdom of Italy was proclaimed in 1861 the country was fragmented, consisting of the Kingdom of Sardinia (Piedmont, Sardinia, Savoy and Genoa), the Kingdom of the Two Sicilies (which included Naples and Sicily) and Tuscany. The pope was also returned to his estates. The area of Lombardy and Venetia to the top right of the detail to the left was under the rule of the Habsburgs, establishing Austria as the dominant influence in the region although a series of minor duchies also existed in central northern Italy.

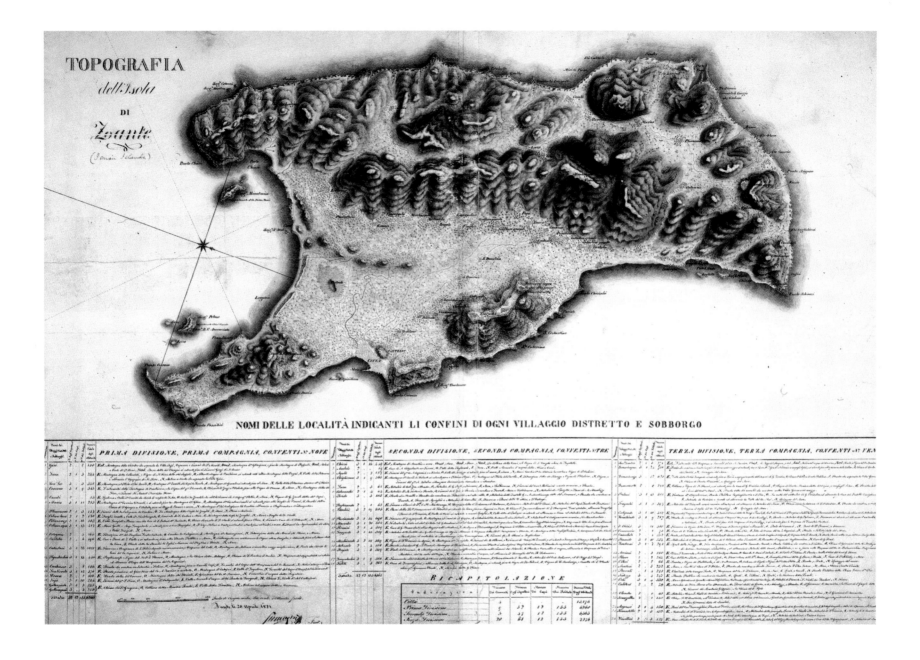

118

Greece, Albania 1827

Left: This 1927 map boasted that it was 'A New Map of Greece Constructed chiefly from original materials, in which it has been attempted to improve the Ancient and Modern Geography of that Country'. It shows rivers, roads, towns, hachures (lines that indicate topographical relief) and archaeological sites. In 1827, Britain, Russia and France signed an agreement to force Turkey to make an armistice with the Greeks and to grant them autonomy. This led to the Battle of Navarino where the joint British, French and Russian squadrons destroyed the Turkish and Egyptian fleets. This catastrophe to Turkey rendered the independence of Greece inevitable. From the Crimean War onwards, the successes of nationalists in fighting Ottoman rule never flagged and in 1913, in the last war of the Balkan states against Turkey, an autonomous Albania emerged.

Zante 1821

Far Left: Titled 'Topografia dell'Isola di Zante (Ionian Islands)', this map features a table of inspectore, 'capi', soldiers and inhabitants of each village and the city of Zante itself. The island of Zante (the Venetian name; the Greeks call it Zakynthos) is the sixth of seven Ionian isles renowned for their lush beauty. It was an important stronghold of the Venetians from the early 17th century (while the rest of Greece was under the Ottoman Turkish rule) until the extinction of their republic in 1797. Zante was awarded to the British at the Congress of Vienna and remained in British hands until it became part of independant Greece in 1862. To this day the island is known as 'fiore to lavante' — flower of the east.

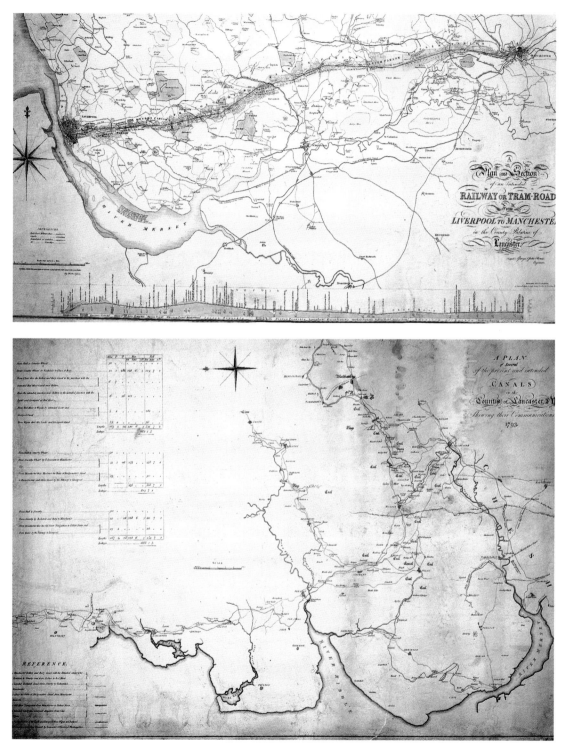

Liverpool–Manchester Railway c.1826

Left: Showing a route from the busy port of Liverpool to Manchester, the centre of Britain's industrialised north, this map plans a pioneering rail or tram road that would take raw materials such as cotton from America to the factories of Manchester and the Lancashire textile industries. At the time steam locomotion was in its infancy; the engine that won a competition to find a suitable locomotive for the new track — the famous Rainhill trials — was the *Rocket*, built by George and Robert Stephenson between 1826 and 1829.

Canals of Lancashire c.1826

Below Left: Stretching from the ports of Milnthorp, Lancaster and Preston through Bolton to Manchester, this map shows the existing canals of Lancashire as well as those that were planned to facilitate transportation between the ports and the great factories. The canal marked in blue, which starts at Liverpool and ends at Colne, via Wigan, would eventually go on to Leeds in West Yorkshire, a distance of some 120 miles. A table in the top left hand corner also gives information about the distance between the wharves and the average tidal levels.

Lancashire c.1820

Right: The County of Lancaster has always been an important part of Britain's industrial might with major towns in the form of Liverpool and Manchester as well as the substantial coal deposits, which powered the machinery of the Industrial Revolution. At the time it was drawn, this map represented a populous and heavily industrialised region of Britain, whose ports and shipowners had grown wealthy on trade — particularly slaving. Traditionally, however, the county's biggest product was cotton textiles; indeed, it was the world's largest producer for many years. The map is dedicated to the Lord Lieutenant of Lancashire and has abundant notes about the history of the county and its towns.

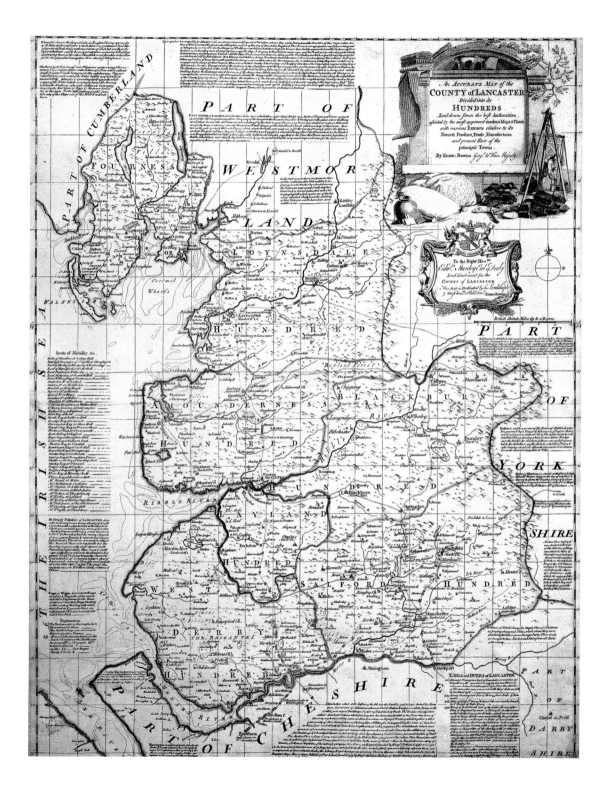

An ACCURATE MAP of the
COUNTY of LANCASTER
Divided into its
HUNDREDS
Laid down from the best Authorities,
afsisted by the most approved modern Maps & Charts
with various Extracts relative to its
Natural Produce, Trade, Manufactures
and present State of the
principal Towns.

By Eman. Bowen Geogr. to His Majesty.

To the Right Hon.ble
Edwd. Stanley Earl of Derby
Lord Lieutenant for the
COUNTY of LANCASTER
This Map is Dedicated by his Lordships
Most humble & Obedt. Servt. Eman. Bowen

Montenegro 1856–78

Left: The Congress of Berlin was held from 13 June to 13 July 1878, and was attended by every European power. The primary objective of the congress was to revise the earlier San Stefano Treaty, dividing the Balkan states and halting Russian expansion in the region. Only three states — Serbia, Montenegro and Romania — were allowed their independence (though all were denied the territories that they were eager for). Elsewhere, the Balkans were dished out arbitrarily, with Austria taking Bosnia and Herzegovina, while Britain received Cyprus and Russia Bessarabia. As shown in the key, the result of this cynical power-broking was that the state of Montenegro became significantly smaller than was agreed by the prior treaty, which had increased its territory considerably. The Balkan nations responded violently to the treaty and it failed completely soon after.

Istanbul 1836

Right: This plan of Istanbul — then still Constantinople — dating from 1836, shows the city as engraved by U. Muschani with an inset showing the Princes Islands. Constantinople had long been of great strategic significance, both militarily and commercially, as it controls the Bosporus, the entrance to the Black Sea, and stands at the traditional crossroads of Europe and Asia. Taken by the Ottomans in 1453, its fall signified both the end of the last vestiges of the Eastern Roman Empire and also the dominance of the Turks. During the period of this map France was active in dismembering Turkey to assist her own Mediterranean schemes.

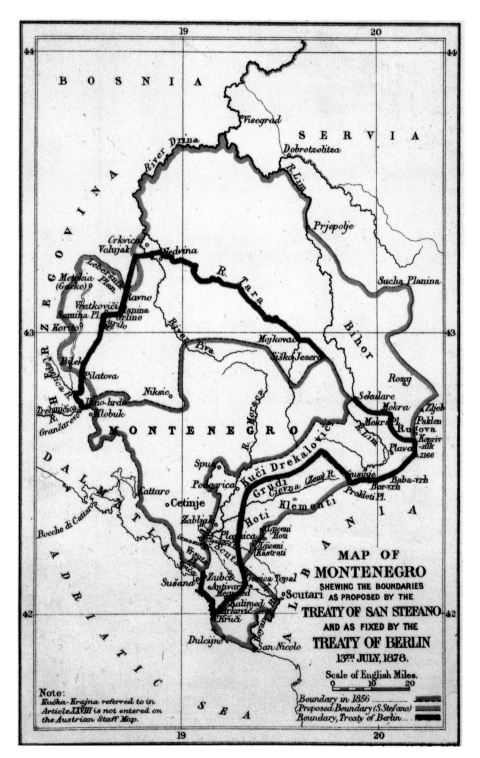

MAP OF MONTENEGRO SHEWING THE BOUNDARIES AS PROPOSED BY THE TREATY OF SAN STEFANO AND AS FIXED BY THE TREATY OF BERLIN 13TH JULY, 1878.

Scale of English Miles.
0   10   20

Boundary in 1856 ....
Proposed Boundary (S.Stefano) ....
Boundary, Treaty of Berlin ....

Note:
Kučka-Krajna referred to in Article XXVIII is not entered on the Austrian Staff Map.

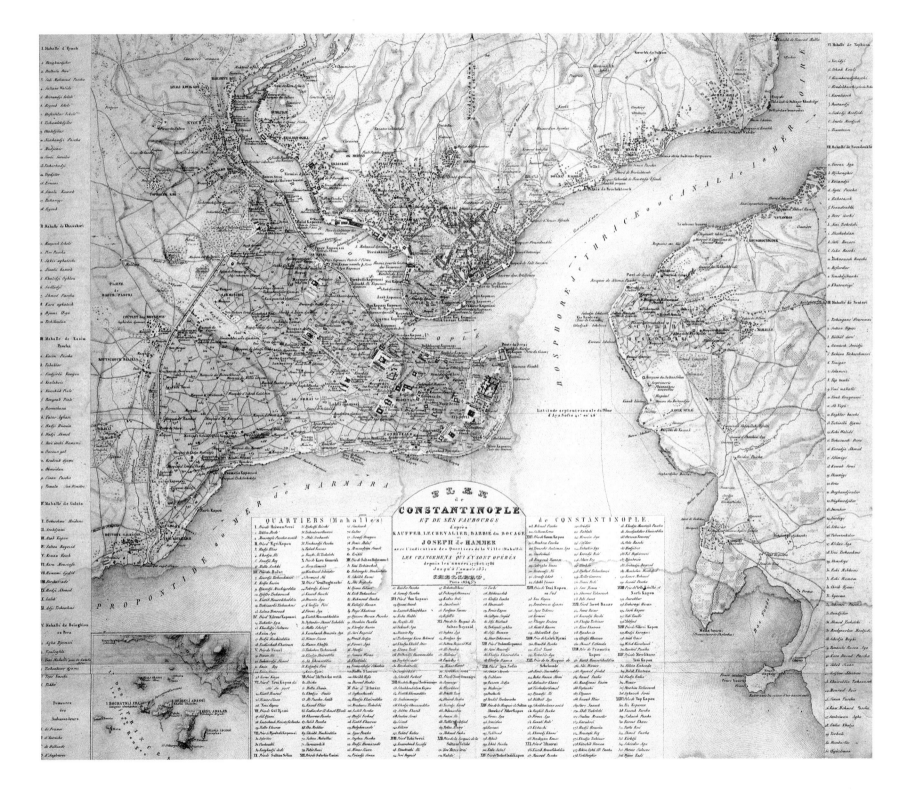

123

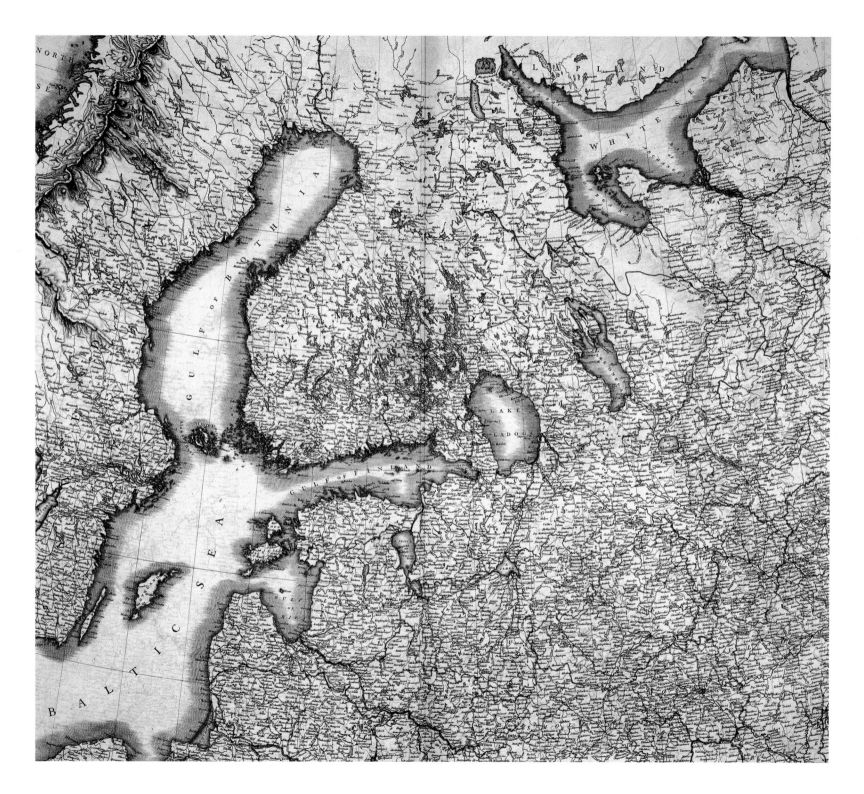

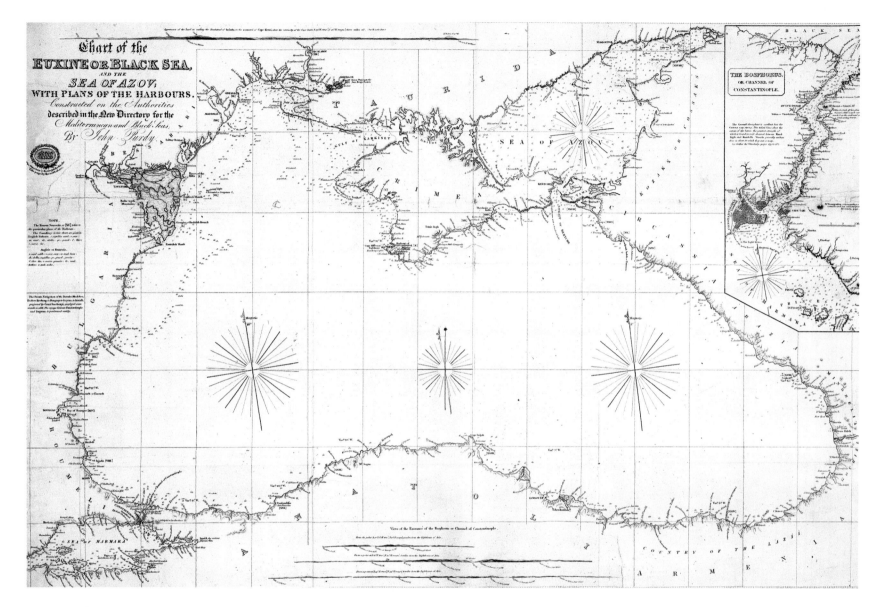

Russia 1828

Left: At the southern end of Lake Ladoga in the centre of this map lies St. Petersburg, at the time the heart of the Russian Empire (Moscow, the medieval capital, became the capital again in 1918). This map was published three years after the Decembrist Revolt threatened to unseat Tsar Nicholas I (1796–1855) from an unstable throne that had been refused by his brother Constantine after the death of Alexander I and was still in dispute. Nicholas overcame these problems with characteristic decisiveness and the remainder of his reign was marked by autocratic repression of his people.

Black Sea 1848

Above: First published by R. H. Laurie of Fleet Street, London, in 1834, this map is an updated edition dating from 1848. It shows the 'Euxine or Black Sea and the Sea of Azov with plans of the harbours'. An area of great strategic and commercial importance, it was controlled by the Turks until the late 18th century when it was gradually wrested from their control by Russia who struggled to maintain dominance in the area. To the east of the Black Sea are the Caucasus Mountains, the scene of many battles between Russia and the Ottoman Empire as Nicholas I attempted to gain supremacy in the region.

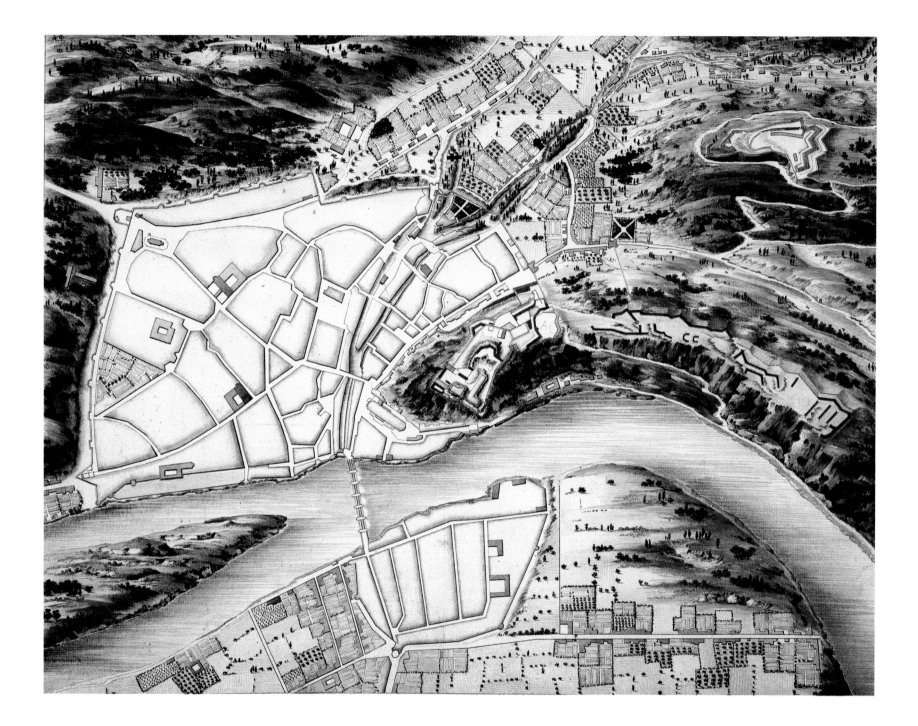

Koblenz 1834

Left: This map shows the confluence of the rivers Rhine and Mosel at Koblenz. As with so many areas where two rivers converge, the region is of great strategic importance and castles and bastions have been built, including both the Citadel of Ehrenbreitstein and Fort Alexander. This map was executed by the British military and shows the fortifications as well as the surrounding countryside with its plateaux, ravines and the hillside behind the fortress.

Edinburgh 1837

Above Right: Edinburgh has been a European cultural and intellectual centre since its university — the largest in Scotland — opened its doors in the 16th century. At the beginning of the 19th century it was decided by the city council that, if the city was to prosper, it must grow. A New Town, an ideal city laid out rationally, was to be built. This project was entrusted to a little known architect, James Craig, whose grand plan was to build a town to equal London in stature.

Edinburgh Castle 1836

Right: The first wooden fortress was built in Edinburgh by Malcolm III during the 11th century and since then the castle has evolved through many additions and restructurings into an imposing edifice looming over the city from its rocky perch. Built on an extinct volcano, the castle is at one end of the 'Royal Mile' that leads to the more comfortable royal palace of Holyroodhouse. This plan details the castle's many buildings, with a key giving information on each.

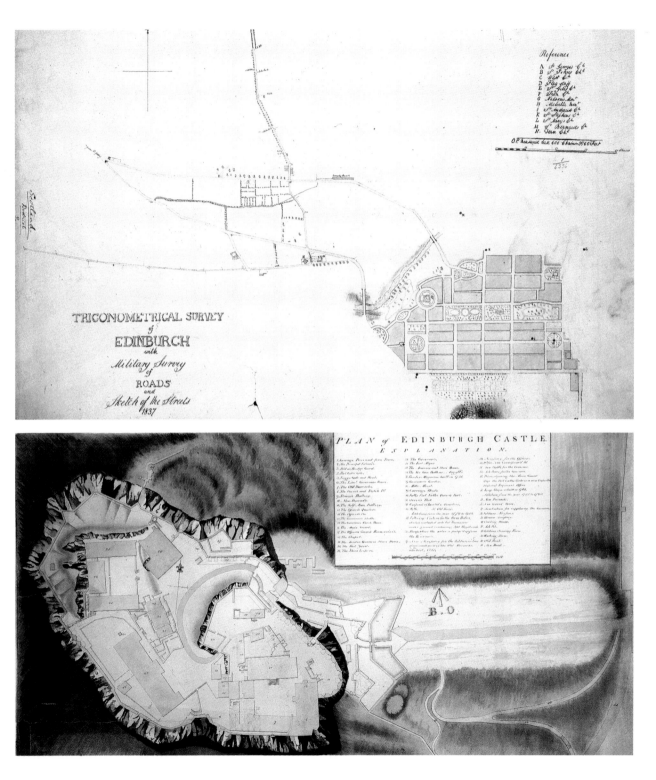

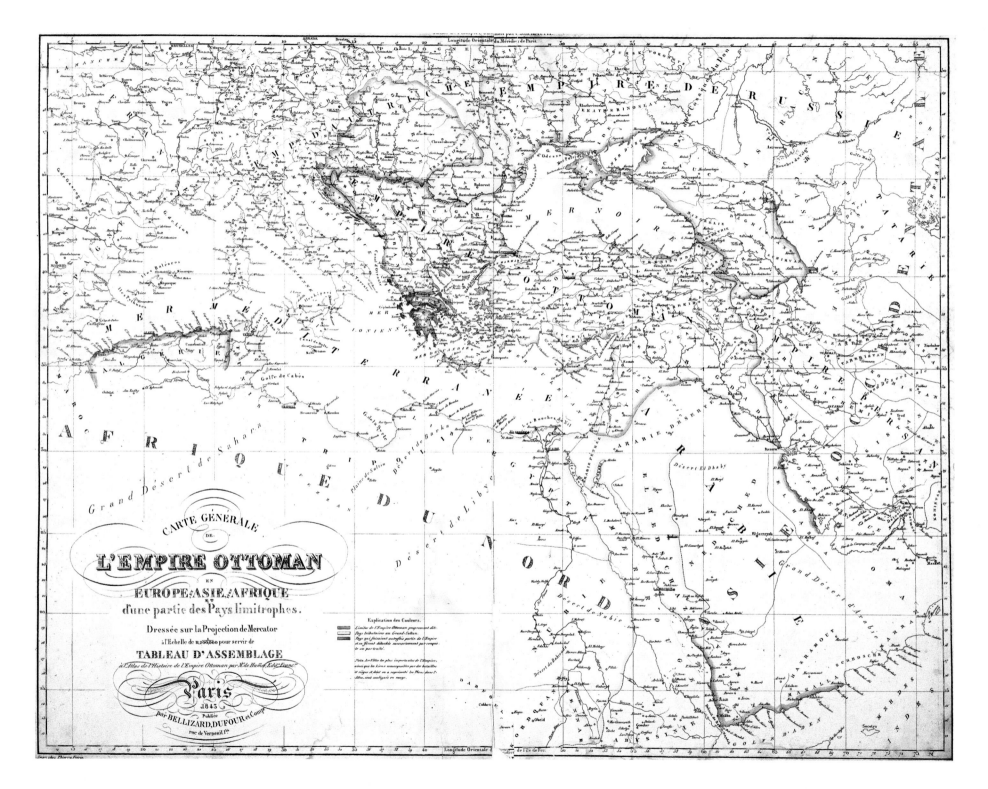

128

Ottoman Empire 1843

Left: By the middle of the 19th century the Ottoman Empire, which had ruled over south-eastern Europe, the Arab Middle East and North Africa for centuries, was declining rapidly. Internal politics caused massive rifts within the empire as traditionalist and progressive factions threatened the nation's security and a series of conflicts with Russia and Austria eroded territory. Known as the 'Sick Man of Europe', the Empire managed to endure the Crimean War of 1853–56, with military aid from Britain, but declined further until it was finally broken up after World War I.

Hungary c.1845

Right: The mid-19th century also marked over a century of Austrian Habsburg rule in Hungary. However, during the early part of the century a nationalist movement, initially interested in promoting Hungarian culture, gained momentum and became a political party. In 1848, a War of Liberation against Austria (which was supported by Russia) broke out but ended in defeat for Hungarian leader Lajos Kossuth three years later. Continued resistance from the Hungarian population together with political shifts throughout Europe nevertheless eventually led to the Austro–Hungarian Compromise of 1867, which forged a dual monarchy for the two countries.

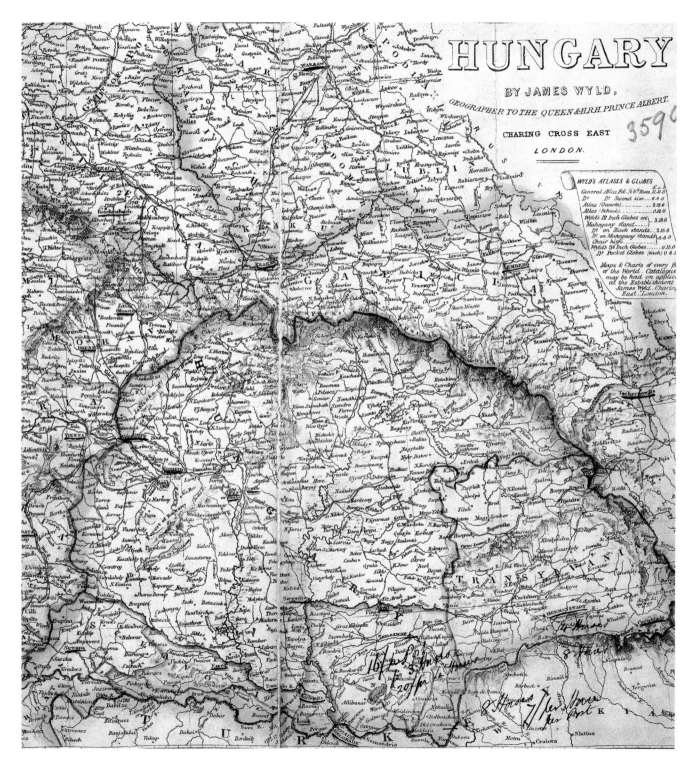

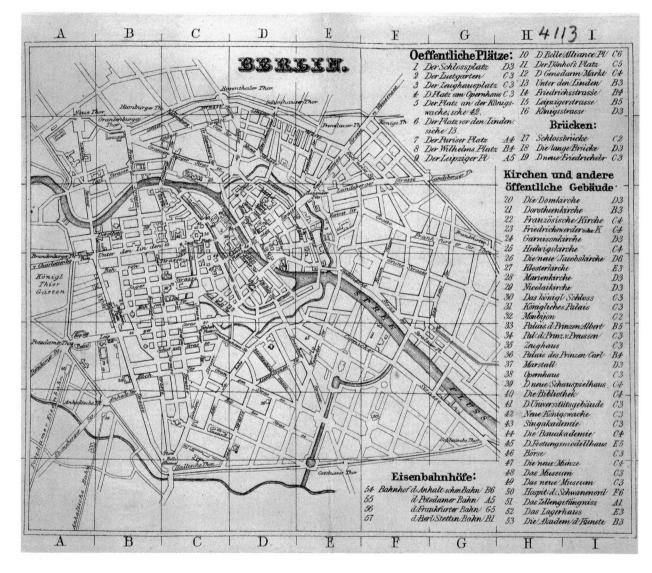

BERLIN.

Oeffentliche Plätze:

1	Der Schlossplatz	D3
2	Der Lustgarten	C3
3	Der Zeughausplatz	C3
4	D.Platz am Opernhaus	C3
5	Der Platz an der Königs- wache, sche 42.	C3
6	Der Platz vor den Linden siche 13.	
7	Der Pariser Platz	A4
8	Der Wilhelms Platz	B4
9	Der Leipziger Pl.	A5
10	D Belle Alliance Pl.	C6
11	Der Dönhof's Platz	C5
12	D Gensdarm Markt	C4
13	Unter den Linden	B3
14	Friedrichsstrasse	B4
15	Leipzigerstrasse	B5
16	Königsstrasse	D3

Brücken:

17	Schlossbrücke	C2
18	Die lange Brücke	D3
19	D neue Friedrichsbr.	C3

Kirchen und andere öffentliche Gebäude:

20	Die Domkirche	D3
21	Dorotheenkirche	B3
22	Französische Kirche	C4
23	Friedrichswerder sche K	C4
24	Garnisonkirche	D3
25	Hedwigskirche	C4
26	Die neue Jacobskirche	D6
27	Klosterkirche	E3
28	Marienkirche	D3
29	Nicolaikirche	D3
30	Das königl Schloss	C3
31	Königliches Palais	C3
32	Monbijou	C2
33	Palais d. Prinzen Albert	B5
34	Pal. d. Prinz. v. Preussen	C3
35	Zeughaus	C3
36	Palais des Prinzen Carl	B4
37	Marstall	D3
38	Opernhaus	C3
39	D neue Schauspielhaus	C4
40	Die Bibliothek	C4
41	D.Universitätsgebäude	C3
42	Neue Königswache	C3
43	Singakademie	C3
44	Die Bauakademie	C3
45	D.Festungsmodellhaus	E5
46	Börse	C3
47	Die neue Münze	C4
48	Das Museum	C3
49	Das neue Museum	C3
50	Hospit. d. Schwanenord	F6
51	Das Zellengefängniss	A1
52	Das Lagerhaus	E3
53	Die Akadem d. Künste	B3

Eisenbahnhöfe:

54	Bahnhof d Anhalt schen Bahn	B6
55	d. Potsdamer Bahn	A5
56	d. Frankfurter Bahn	G5
57	d. Berl. Stettin Bahn	B1

Berlin c.1850

Left: Published by Bradshaw & Blacklock of Manchester, this litho-graphed plan of Berlin includes a reference table showing waterways, bridges, streets, public places, build-ings, churches, gates and railway stations (in Germany, as in the rest of Europe, there was a huge growth of the railway in the 19th century). The city, situated on the River Spree, traditionally boasted a large popula-tion of artists and intellectuals and was richly endowed with magnificent architecture as well as a large popu-lation, all of which earned it the nick-name of 'the Athens on the Spree'. In 1871 it became the capital of unified Germany.

Naples c.1850

Right: This street map of Naples showing surrounding hills and fortifi-cations is inset at the bottom right with a harbour view showing Mount Vesuvius, which has caused devasta-tion to the region many times over the centuries. The city itself has a long-standing attraction for artists and is blessed with fine architecture including its five medieval castles as well as over 200 churches. It was as popular a destination for Victorian travellers as it is today. At the time this map was made Naples was the heart of the independent Kingdom of Naples, which would survive until the unification of Italy in 1861.

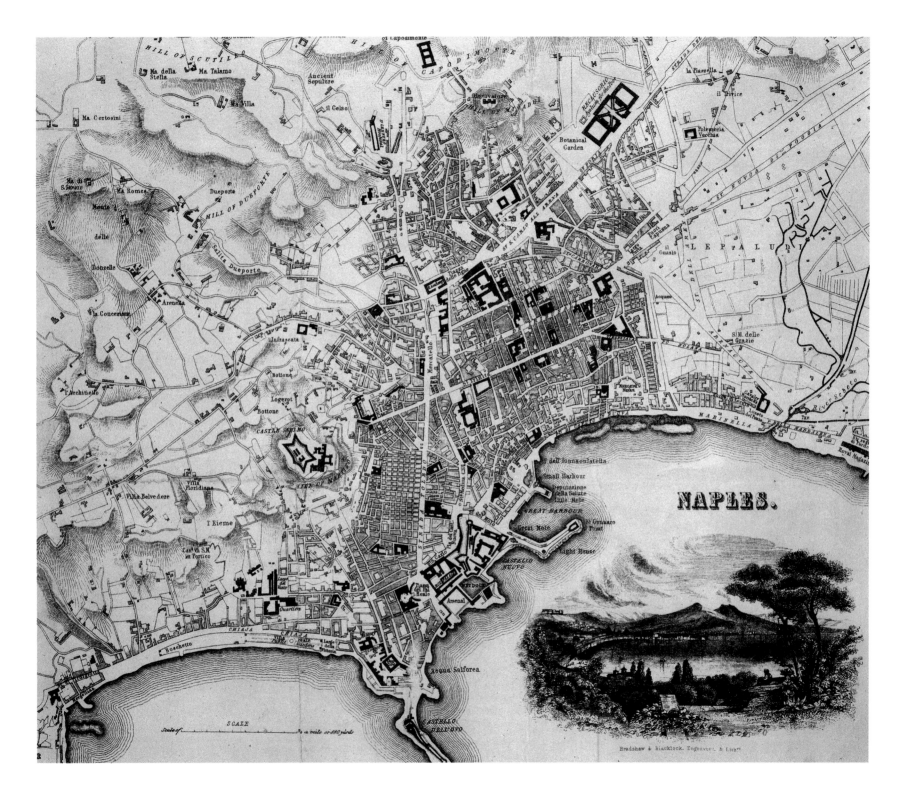

NAPLES.

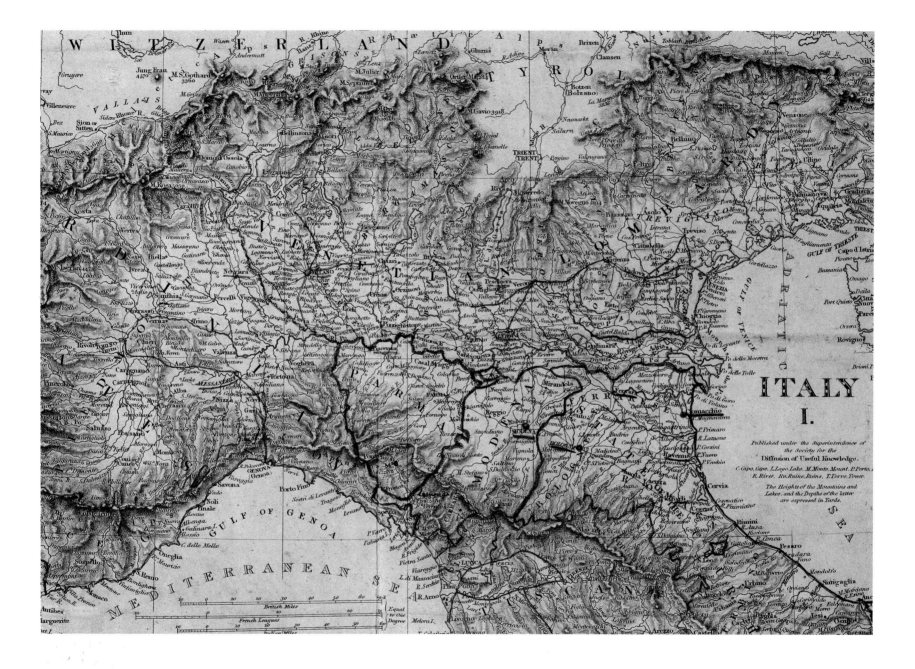

ITALY

I.

Published under the Superintendence of
the Society for the
Diffusion of Useful Knowledge.

C. Capo, Cape. L.Lago, Lake. M. Monte, Mount. P. Porto,
R. River. Ru. Ruine, Ruins. T. Torre, Tower.

The Heights of the Mountains and
Lakes, and the Depths of the latter
are expressed in Yards.

132

Italy 1853

Left: This map, published in Britain on 1 January 1853 under the supervision of the philanthropic Society for the Diffusion of Useful Knowledge, is a beautifully rendered chart of northern Italy. By 1853 this region was still recovering from the revolts of 1848, during which the king of Sardinia-Piedmont, Charles Albert (1798–1849), declared war on Austria with the goal of ousting the latter from northern Italy. He was beaten, however, and the Habsburgs were restored to their lands.

Finland 1854–60

Below and Detail Right: These two maps, printed by the Military Topographical Depot, St. Petersburg, in 1860, were compiled from the military reconnaissance of 1854–55 by Colonel Alftan, General Staff. The maps include a reference table to lakes and river systems and an inset of the city of Helsinki.

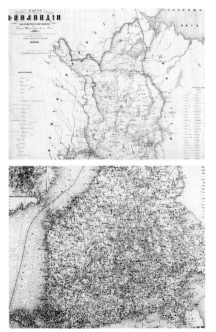

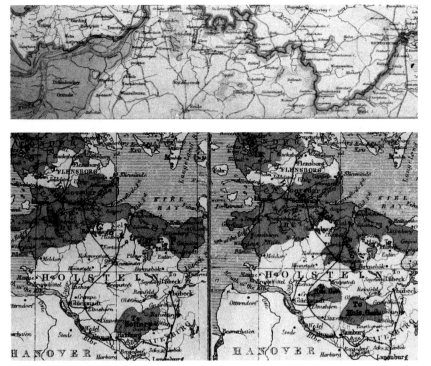

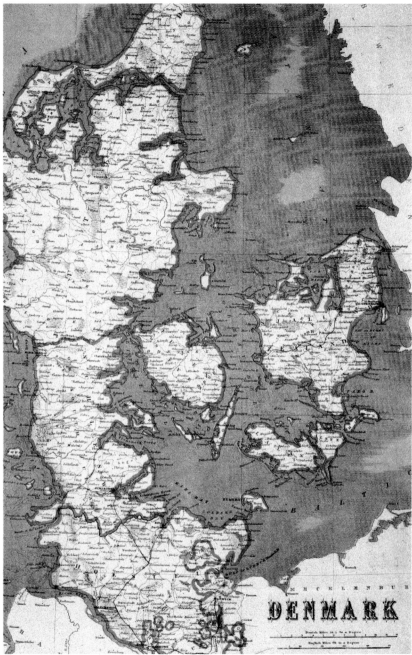

Denmark 1864

Above, Right and Detail Far Right: Long-standing differences between the two duchies of Schleswig and Holstein and Denmark, over royal succession, resulted in the former taking up arms against Denmark during the revolutionary outbreak of 1848–49. European forces swiftly built up — Prussia on the side of Schleswig, which had German sympathies, and Austria supporting Denmark and Holstein. After a number of inconclusive battles Prussia and Austria turned to arbitration and the question was finally settled without further bloodshed. The right of succession was awarded, by the Protocol of London, 8 May 1852, to Prince Christian of Schleswig-Holstein-Sonderburg-Glucksburg and his male descendants. The protocol was, however, violated by King Frederick VII of Denmark in 1863 causing Prussian forces to invade Holstein. In early 1864 Prussia formed an alliance with Austria and both countries entered Schleswig. The overwhelmed duchies were surrendered to the Austro–Prussian alliance by the Treaty of Vienna in October 1864 with Prussia taking possession of Schleswig and Austria administering Holstein.

This map, made after the conflict, details Denmark itself with the duchies (**Above**) and the course of the River Eider and its defences (**Top**) as detailed inserts. To the left of the map (not shown) is a history of the region that explains the circumstances of the dispute and its eventual outcome.

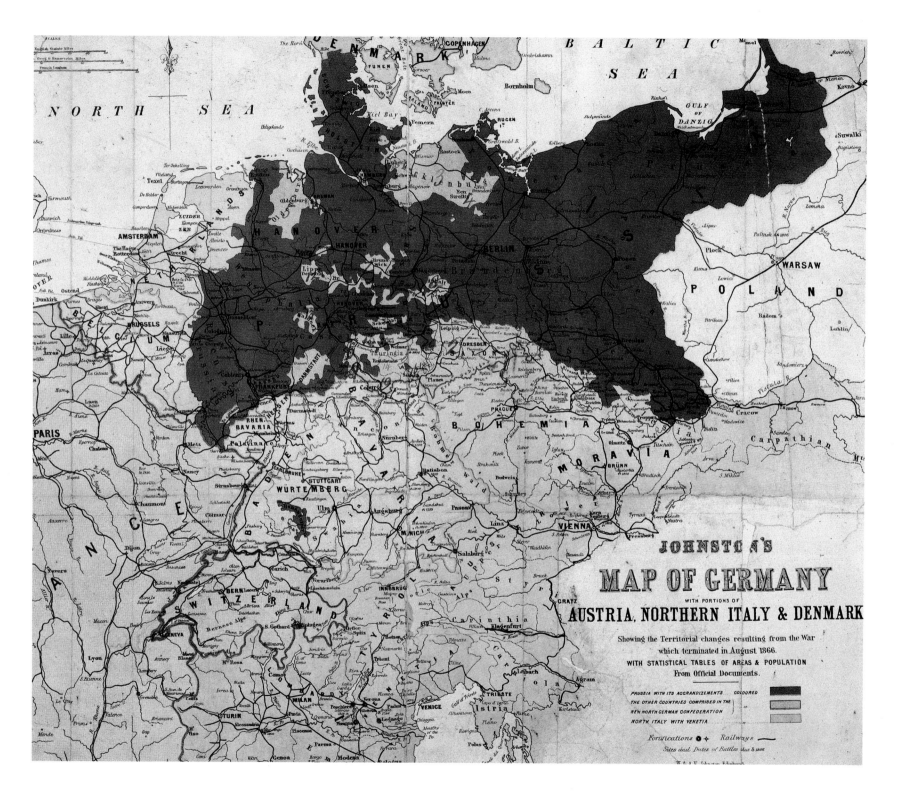

JOHNSTON'S
MAP OF GERMANY
WITH PORTIONS OF
AUSTRIA, NORTHERN ITALY & DENMARK

Showing the Territorial changes resulting from the War
which terminated in August 1866.
WITH STATISTICAL TABLES OF AREAS & POPULATION
From Official Documents.

PRUSSIA WITH ITS AGGRANDIZEMENTS   COLOURED

THE OTHER COUNTRIES COMPRISED IN THE
NEW NORTH GERMAN CONFEDERATION

NORTH ITALY WITH VENETIA

Fortifications   Railways
Sites and Dates of Battles

Central Europe 1866–67

Left: The Austro–Prussian War, also known as the Seven Weeks' War, broke out in 1866 following furious power-brokering in the aftermath of the Danish defeat of 1864. The short war, lasting only six weeks, was an easy victory for Prussian forces. The subsequent Peace of Nikolsburg saw Austria cede Venetia to Italy but, anxious to pursue friendly relations in the future, Prussia did not press for further reparations or any other territories. The principal outcome, as shown in the map, was the division of Germany into the North German Confederation, Bavaria, Württemberg and Baden.

Central Europe 1866

Right: 'Stanford's general map of the Germanic Confederation, also Schleswig Holstein & Venetia'. Under the terms of the Treaty of Prague, signed in 1866, Austria was excluded from German Affairs and forced to cede Venetia to Italy. As discussed above, the German Confederation of 1815 was dissolved with Schleswig-Holstein, Hanover, the whole of Hesse-Cassel, Nassau and the free city of Frankfurt annexed by Prussia. The North German Federation was formed comprising Prussia, Saxony, the grand duchies of Oldenburg, the two Mecklenburgs, Brunswick, part of Hesse Darmstadt, the Thuringian States and the free cities of Hamburg, Bremen and Lübeck. An understanding was reached between Prussia and the south German states in the form of an offensive and defensive alliance based on a reciprocal guarantee of territorial integrity.

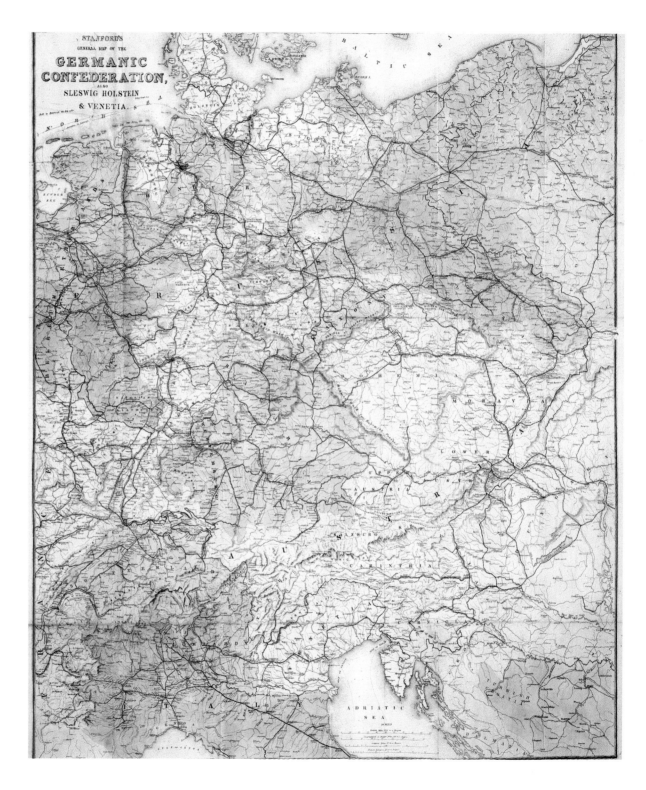

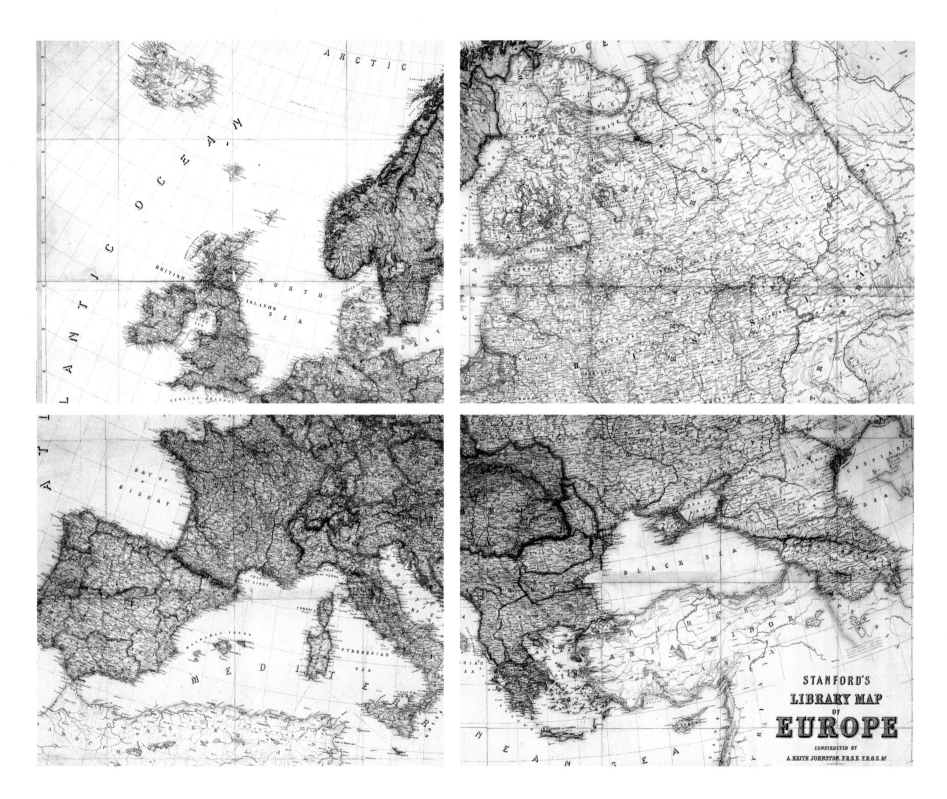

STANFORD'S
LIBRARY MAP
OF
EUROPE
CONSTRUCTED BY
A. KEITH JOHNSTON, F.R.S.E. F.R.G.S. M.

Europe 1858

Left and Detail Right: This series of four maps make an interesting comparison with a similar set on page 110, made just over 60 years earlier. 'Stanford's Library Map of Europe constructed by A. Keith Johnston, F.R.S.E., F.R.G.S. &c' was published by Edward Stanford, London, on 1 June 1858. It shows highlands, rivers, canals, roads, railways, submarine and telegraph lines, and boundaries. It is still too early to recognise today's political boundaries from this map: Germany and Italy have yet to be unified and the Turkish Empire still thrusts halfway up the Balkans into divided Croatia.

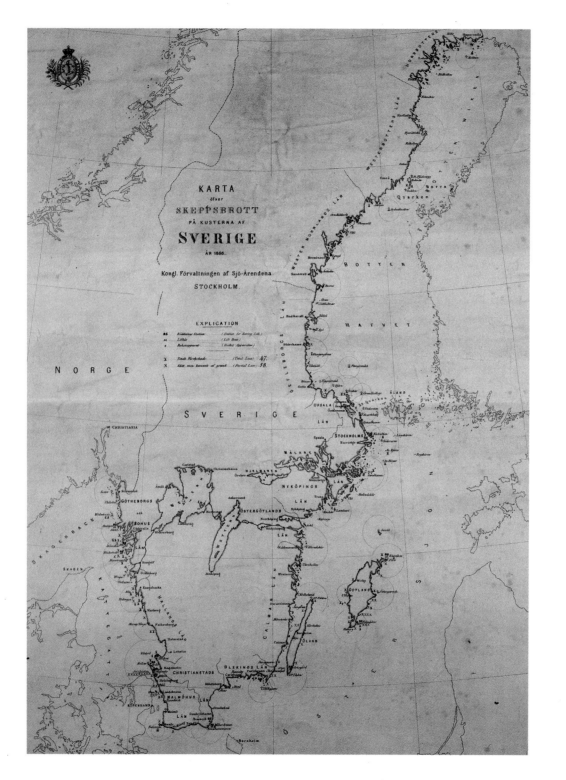

Sweden 1866

Left: Lithographed by W. Schlacter, this map of Sweden was published by the Swedish Admiralty, Stockholm. It shows lifeboat stations, lifeboats, rocket apparatus and wrecks with total — 47 — or partial — 58 — loss of life. The bulk of the accidents took place, as one might expect, in the dangerous waters between Sweden and Denmark, the Skagerrak, Kattegat and the narrows between Helsingborg and Copenhagen.

Corinth Canal 1893

Right: This French drawing provides seven plans, views and sections of the Corinth Canal and includes maps showing the improvements it afforded to shipping of the period by reducing journey time and distance. The canal is short — just over six kilometres — and was achieved by cutting through some 86m of rock to reach sea level. The drawings are dated 1893, the year the canal was opened.

CANAL MARITIME DE CORINTHE

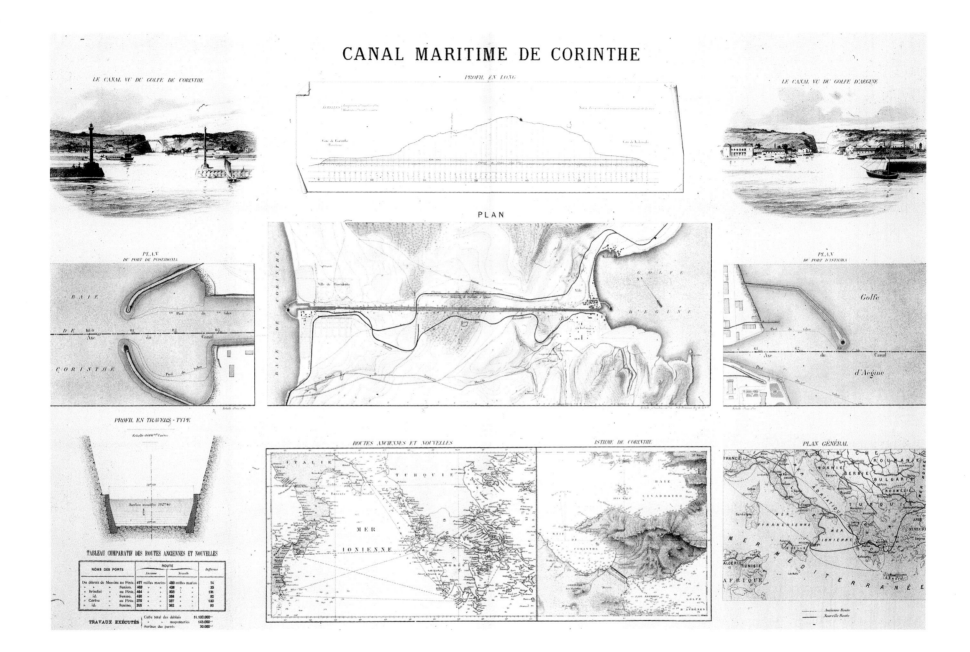

Balkans 1897

Left: Printed and published by Stanford's Geographical Establishment, London, this 'Map to illustrate notes on Races and Religions in Turkey-in-Europe' illustrates the states and provinces of the Balkan Peninsula with notes on races and religion of Turkey-in-Europe. The map indicates the various religions which predominated in the separate Balkan states — something that hasn't changed much through to today. The key to the colours on the map shows the Turks in pink, Albanian Catholics in light green, Albanian Moslems in dark green, Bulgarian Christians in light yellow, Bulgarian Moslems in dark yellow, Serbian Christians in light red, Serbian Moslems in dark red and Greeks in blue.

Channel Islands

Right: A chart of the islands of Jersey, Guernsey, Sark, Herm and Alderney with the adjacent coast of France.

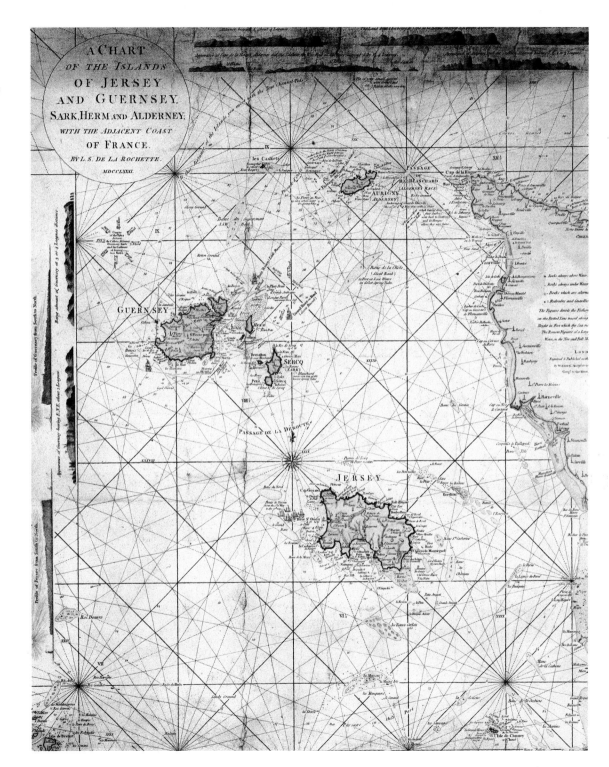

INDEX OF MAPS

Aragon 1719 — 47
Austrian Netherlands 1789 — 80

Balkans 1897 — 142
Battle of Blenheim 1704 — 39
Battle of Fleurus 1690 — 33
Battle of Mohács 1526 — 24
Belgium 1814 — 106
Belgium and the Netherlands 1701–12 — 38
Belgrade 1717 — 45
Berlin 1757 — 62
Berlin c.1850 — 130
Black Sea 1848 — 125
Brussels 1745 — 59

Cadiz Harbour 1587 — 27
Canals of Lancashire c.1826 — 120
Central Europe 1763 — 66
Central Europe 1866–67 — 136
Channel Islands — 143
Copenhagen 1807 — 104
Corfu 1716 — 44
Corinth Canal 1893 — 141

Denmark 1864 — 134
Dutch Coast 1673 — 31

Eastern Lombardy 1736 — 50
Edinburgh 1837 — 127
Edinburgh Castle 1836 — 127
Europe 1783–1807 — 74
Europe 1800 — 90
Europe 1805 — 102
Europe 1815/16 — 110
Europe 1858 — 138
European Courier Routes c.1820 — 114

Finland 1743 — 56
Finland 1854–60 — 133
France 1756 — 63
France 1790–91 — 83
France 1815/16 — 113

German Empire 1715 — 42
Ghent Canal 1696 — 37
Greece, Albania 1827 — 119

Home Counties, England 1774–76 — 70
Hungary c.1845 — 129

Île de Ré 1700 — 36
Ireland 1567 — 25
Istanbul 1836 — 123
Italy 1803 — 88
Italy 1825 — 116
Italy 1853 — 132

Kingdom of the Two Sicilies 1783 — 73
Koblenz 1834 — 126

La Rochelle 1718 — 46
Lancashire c.1820 — 121
Liverpool–Manchester Railway c.1826 — 120
Low Countries 1730 — 49
Luxembourg City 1801 — 89

Madrid 1785 — 78
Mediterranean Sea 1738 — 54
Montenegro 1856–78 — 122

Naples c.1850 — 131
North Italy — 103
Norway and Sweden 1785 — 79

Ottoman Empire 1843 — 128

Partition of Poland 1658–1815 — 68
Pas de Calais 1707 — 40
Peninsular War 1808–14 — 109
Portugal 1653 — 30
Portugal 1790 — 82
Prague 1742 — 55
Pyrenees–Adriatic 1811 — 105

River Charente 1775 — 71
Rome 1748 — 60
Russia 1815/16 — 111
Russia 1828 — 124

Scotland 1745/46 — 58
Siege of Philipsbourg 1734 — 52
Siege of Vienna 1529 — 24
Spain 1770 — 108
St. Malo 1693 — 34
St. Malo 1700 — 35
Sweden 1866 — 140
Switzerland 1782 — 72

Tangier 1679 — 32
The Rhine 1711 — 41
Toulouse 1720 — 48

Ukraine 1736 — 53
United Kingdom 1815/16 — 112

Wales 1610 — 28
Western France 1636 — 29
Westminster 1585 — 26

Zante 1821 — 118